Barcode in Back

MW01122101

HUMBER LIBRARIES LAKESHORE CAMPUS
3199 Lakeshore Blvd West
TORONTO, ON.   M8V 1K8

MAGIC PACKAGING

DESIGNERBOOKS

MagicPackaging

Copyright © 2010 DESIGNERBOOKS

Organiger: Jeff Li
Editor: Wu Jun, Wei Xiaojuan, Jeff Li
Design and Layout: Jeff Li
Proofreader: Cheng Ping

Publisher: Chen Zhong
DESIGNERBOOKS
Unit D, 16/F, Cheuk Nang 21st Century Plaza, 250 Hennessy Road,
Wanchai, Hong Kong
Tel: +852-2575-5186
Fax: +852-2891-1996
E-mail: edit@designerbooks.com.cn

Distributor:
DESIGNERBOOKS
B-0619, No.2 Building, Dacheng International Center, 78 East 4th Ring Middle Road,
Chaoyang District, Beijing, China
Tel: 0086-10-5883-1335 (Beijing)   0086-22-2341-1250 (Tianjin)
     0086-21-5596-7639 (Shanghai)   0086-571-8884-8576 (Hangzhou)
     0086-25-5807-5096 (Nanjing)   0086-23-6772-5751 (Chongqing)
     0086-755-8825-0425 (Shenzhen)   0086-20-8756-5010 (Guangzhou)
     0086-28-8665-0016 (Chengdu)   0086-27-6566-2067 (Wuhan)
Fax: 0086-10-5962-6193
E-mail: info@designerbooks.com.cn
http://www.designerbooks.com.cn

Printed in China

All rights reserved. No part of this publication may be reproduced in any form or
by any means, graphic, electronic or mechanical, including photocopying and
recording by an information storage and retrieval system without permission in
writing from the publisher.

ISBN 978-988-18078-9-2

CONTENTS

Delicious Magic

006-140

Intriguing Magic

142-216

Dizzying Magic

218-267

Delicious Magic

Milk packaging design

This hungarian brand produce high quality dairy products, but the look of those package is so clumsy and old fashioned. We'd like to make a new and powerful design that is totally different from the other competitors. Clear-out, colour coded packs with minimal style typography. In a simple, but progressive look.

Client
Jasztej Dairy Company

Design Agency
Fontos Graphic Design Studio (Hungary)

Creative Director
Mate Olah

Designer
Mate Olah

Other
3D modelling by Gabor Gloviczki

Concept redesign "Mom's Healthy Secrets"

Mom's Healthy Secrets is a brand of cereals distributed across Canada. The Package was inspired by the typical Milk and Fruit-Boxes. Both food fits perfect to a yummy Muesli meal. The Package Design is pure-handmade fonts (Mom's heritage) combined with separate ingredient to enhance a modern and unseen approach.

Client
Mom's Healthy Secrets, Canada

Design Agency
TIBOR+ (Germany)

Creative Director
Tibor Hegedues

Designer
Tibor Hegedues

Photography
Stockmaterial

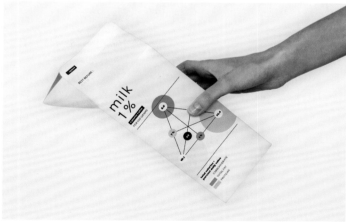

Client
self-initiated

Design Agency
FFunction (Canada)

Creative Director
Audree Lapierre

Designer
Audree Lapierre

Nutritional Data Packaging

Our concept was to design a packaging using nutritional facts about the food product. For the milk carton, we used the four sides to inform about the caloric ratio, nutrience balance completeness, ingredients and amount per serving.
The data visualizations say more than a regular nutritional-facts label. For example, ingredients are visually linked to their corresponding components (carbohydrate, total fat, proteins, sodium, others). Nutritional information becomes the main goal of the packaging, while still carrying a powerful branding by combining an expressive form with useful information.

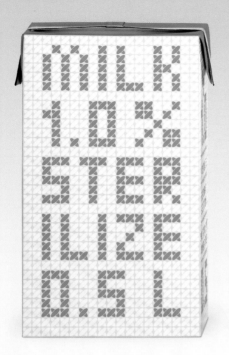
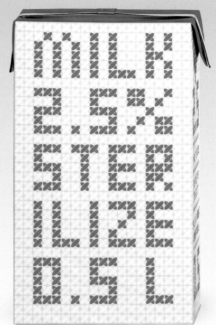

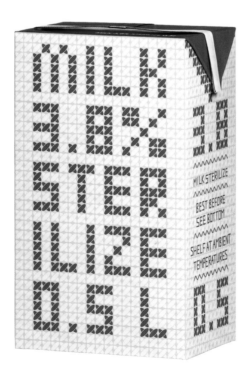

Crestmilk / Milk collection

Hattomonkey design studio has created a brand new dairy beverage product package concept.
So the task which has been set before designers is to invent a package, which is to underline naturality, thereby to get closer to the consumers of diary products. The idea was born unexpectedly—to stylize package by a cross stitch.

Client
Milk collection

Design Agency
Hattomonkey (Russia)

Creative Director
Alexey Kurchin

Designer
Alexey Kurchin

Milkitty

Hattomonkey has created the brand mark and packaging design of milk "Molokoshka". The brand name is based on clench of two Russian words: Moloko (milk) and Koshka (cat) – the greatest milk lover. So the main element of packaging design is a cat that chooses the product because of its high quality.

Client
Molokoshka

Design Agency
Hattomonkey (Russia)

Creative Director
Alexey Kurchin

Designer
Alexey Kurchin

"Soy mamelle" soya milk

The soya milk produced under "Soymamelle" brand is 100% vegetative product with no cholesterol. It is the source of protein and calcium. It is intended for people who are intolerant of lactose included in nature cow's milk.

Unique form of the package like a cow's udder means that milk of vegetative components is identical to the cow's milk. Special tactile feeling that appears when you touch the package tells about product's naturalness as if consumer milks "vegetative udder" himself. It means that the way from a plant to supermarket's shelf is minimal. Such interaction process (interaction between the consumer and the package) clears away a complicated plan of soya milk production (pasteurization, filtration, enrichment and others) of consumer's view. Thereby it equates with natural product.

Client
PDA and IPACK-IMA

Design Agency
KIAN brand agency (Russia)

Creative Director
Kirill Konstantinov

Design Director
Maria Sypko

Designer
Evgeny Mogalev

Other
Bronze Epica awards 2009

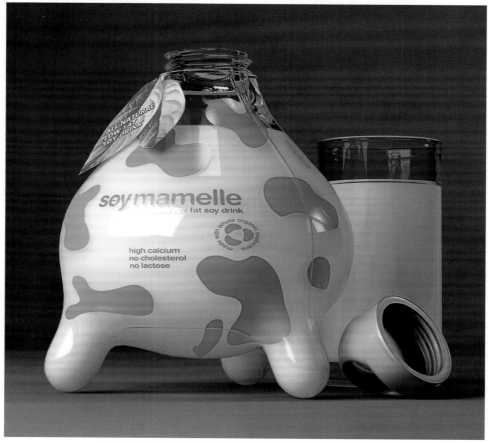

Good Milk

This concept offers the possible broadening of the technology of producing the Tetrapack packaging. The idea includes the use of Congreve stamping in regular milk packages. The design is based on the contrast between the color of the logo and the predominant white, which, as we believe, characterizes the contents of the package very well. Soft pastel shades may be conveniently used for differentiating between various fat content grades.

Design Agency
StudioIN (Russia)

Design Director
Arthur Schreiber

Designer
Maria Ponomareva

Photography
Pavel Gubin

Bill's milkshakes

Client
Bill's Dairy Farm

Design Agency
ilovedust (UK)

Creative Director
Mark Graham

Design Director
Johnny Winslade

Designer
ilovedust

Photography
n/a

Having been asked by Bill's Dairy Farm in South Africa to create carton graphics for their new range of milkshakes, ilovedust created a distinct design solution. Featuring strawberry, chocolate and vanilla flavours, each individual graphic features its own character and specific colour palate. The fun, candy shop style packaging is designed to appeal to a young target market of 12 – 25 year olds.

These have proved incredibly popular and earned us a place in Top 10 Packaging at The Dieline.

Mew Campaign

Client
Goodserve Co.,Ltd.

Design Agency
Subconscious Co.,Ltd.
(Thailand)

Creative Director
Wirush Eakapapan

Design Director
Somjai Reodecha

Designer
Grich Rungruang

Goodserve came to us with requirements to launch a healthy beverage with extraordinary outcomes. They wanted to be different in competitive market. They wanted unique personality with witty & playful communication. They aimed at college students to young jobbers who were out-going, trendy, intellectual, and ambitious. We came up with the packaging that talked about attitude. 'Half full, half empty' referred to optimism & pessimism attitude, while 'fill up & get going' urged and challenged target group to put their effort into what they wanted to achieve. Not only expressing attitude, the packaging also identified the flavour through easy, friendly, and minimal graphic.

From packaging, we created advertising campaign, including BTS and mupi ads. The objectives were to introduce and create brand awareness for Mew. Big, bold, and straightforward, yet attractive key visuals were applied.

Aluminum one way milk pack.

Aluminum one way milk packaging.
Case for 6 bottles, made of molded egg carton like material. (100% recycled/biodegradable packaging material.)

Design Agency
Baita Design Studio (Brazil)

Creative Director
Helena Baita Bueno

Design Director
Heinz Muller

Designer
Helena Baita Bueno

Photography/Renders
Heinz Muller

Terra dell´Oro

In the wine world, heavy on tradition, replacing the bottle is therefore no easy business. Instead of trying to be second best, mimicking or depicting the bottle and label, Liedgren Design saw other possibilities in a large print area. One field is dedicated to the trademark and vital information, leaving the rest open for happy, informal illustrations. Toning down the seriousness of wine drinking, these packs focus on the essential: relaxed socializing.

Client
Åkesson Vin

Design Agency
Liedgren Design (Sweden)

Designer
Liedgren Design

Illustrator
Maja Sten / Agent Bauer

ELLA Milk Products

We have a meaning that market is overloaded with wrong packaging design especially for low fat dairy products, and that is why we want to design something completely different.
We have chosen derivative style with fine, serif typography and simple colors.

Since the product is mostly consumed by women, the design has elements that remind of women magazines and fashion iconography. With this new and progressive packaging we want to stimulate the consumers expectations and make them more aware of the nature of this product.

Client
Mlekara Subotica

Design Agency
Peter Gregson Studio (Serbia)

Art Director
Jovan Trkulja

Designer
Marijana Zaric

Pucko

Client
Arla Foods AB

Design Agency
Neumeister Strategic Design (Sweden)

Creative Director
Peter Neumeister

Design Director
Peter Neumeister

Designer
Mattias Lindstedt

Production Manager
Helene Mellander Holm

Pucko is a classic chocolate drink brand in Sweden with an iconic glass bottle. Arla Foods wanted to refine the brand identity that it would work on additional packaging types and sizes. A complete identity system that was successfully applied to new packaging types without losing the original look and feel of the brand. Considerable increase in sales indicating a successful transfer into new packaging types.

Fresh'n Fruity

The vibrant design visually articulates the strong brand name 'Fresh'n Fruity' and critically uses the new 'hero' brand logo to deliver the promise of voluptuous fruit. This trademark imagery creates a new world of fruit for Fresh'n Fruity to inhabit and own, halting the on-going design imitation of competitors.

The leafy device serves as an integrated logo and the bursting fruit imagery amongst the vines communicates both the variants and the personality of each range. The on-going response from consumers to date has been: "It makes me want to eat yoghurt".

Client
Fonterra

Design Agency
Dow Design (New Zealand)

Creative Director
Donna McCort

Director
Annie Dow

Designer
Andrew Sparrow

Illustrator
Grant Reed

Arla Jigger

Arla Foods is a Swedish-Danish cooperative and the largest producer of dairy products in Scandinavia. The so-called Jigger tetra-pack with milk and cream is sold as portions in many different variations and markets/languages, resulting in very demanding requirements. The most "difficult" packaging challenge in the Arla family – so as it's been called. A system where the different variations are unique, but also belonging to the same family, was invented. Dots creates togetherness, colours drag them apart. We managed to make a simple and ingenious system that mastered the difficult printing requirements economically and sufficiently. This design concept could be seen throughout Scandinavia, in almost every coffee-shop, restaurant or office. It has truly stood the test of time.

Client
Arla Foods AB

Design Agency
Neumeister Strategic Design
(Sweden)

Creative Director
Peter Neumeister

Designer
Peter Neumeister

Hashi / Hot sauces and spices

There is a huge selection of "Hashi" hot sauces: a very hot "Halapenio", extremely hot "Habanero", unbearable hot sauce "Cayenne pepper", as well as wasabi and mustard. That special look #@$#! was created just as soon as we tried this hot sauce at the studio. Hattomonkey's informal packaging design is hot and exciting. This is not just a simple sauce on a store shelf, it shouts the same language you do "#@$#!".

Client
Interra LTD

Design Agency
Hattomonkey (Russia)

Creative Director
Alexey Kurchin

Designer
Alexey Kurchin

Q SKYR Naturell - Packaging

A brand new packaging design for our client Q-Meieriene! We are proud to introduce Q SKYR Naturell where we made both packaging and web solution in a light and fresh design. You can check out tips on how to make fabulous courses and mix your own favorites with this amazing new product. Go taste it, its in a store nearby!

Client
Q-Meieriene

Design Agency
Anti (Norway)

Art Direction
Kjetil Wold

Graphic Design
Heidi Wilhelmsen

Project Manager
Tine Moe

Hannas Peach Melba

Hannas Peach Melba is a fresh cider with natural flavor of peach and vanilla.

Hannas Peach Melba´s Packaging manages to turn out tasty and summerfresh.

After launch in May 2009, Peach Melba immediately took second place on the Swedish cider sales list and in July the first place.

The cider has sales and distribution in Greece and Spain.

Client
Galatea Spirits

Design Agency
Entire (Sweden)

Creative Director
Mattias Brodén

Designer
Mattias Brodén

Photography
Pelle Lundberg

Arano Juice

Client
Arano

Design Agency
Dow Design (New Zealand)

Creative Director
Donna McCort

Director
Annie Dow

Designer
Donna McCort

Illustrator
Samuel Sakaria

The strategy was to focus on the fruits' origin to promote the products' integrity. The simple illustrations captured a rural simplicity and heritage, while promoting the juice's authenticity. The design captured a premium positioning with the brand back into growth within six weeks.

Client
First Blush Inc.

Design Agency
Ferroconcrete (USA)

Creative Director
Yolanda Santosa

Designer
Yolanda Santosa, Sunjoo Park, Wendy Thai

Photography
Paul Taylor

Other
Luellen Renn, Copywriter;
Ronny Widjaja, Web Developer

First Blush

Made from the wine grape varietals of Cabernet, Chardonnay, Merlot and Syrah, First Blush is a juice that Ferroconcrete has transformed into something much more grown-up. Ferroconcrete designed the brand with a focus on the themes of color, beauty and a wide range of health benefits. Recognizing the juice itself as the soul of the brand, Ferroconcrete allowed it to shine through the transparency of the elegant glass bottles. From the website and packaging, to the print, the brand conveys that this juice is fresh-picked, healthy and truly sophisticated. First Blush is grapes all grown up.

Rustad Water

Rustad Water is pure Norwegian spring water, bottled at source in the heart of Norway. Rustad Water comes in a variety of containers; PET bottles, unique glass bottles and environmentally-friendly cartons. The range has mainly been developed for export markets, such as the USA. When designing the Rustad identity, it was very important to convey the positive values associated with Norway and Norwegian crystal clear spring water. Elements and colours from the Norwegian flag were used to give the right national connection while blue and white symbolise water and purity. The result is a unique range of products with a very strong shelf impact.

Client
Rustad Water AS

Design Agency
Strømme Throndsen Design (Norway)

Creative Director
Morten Throndsen

Design Director
n/a

Designer
Eia Grødal

Photography
n/a

Agua de Caldes de Boí

Caldes de Boí is a water company. Our objective was to design a bottle that would be far distant from the already existing Premium water.

Its shape is based on a traditional profile bottle, with a line that unifies the neck and back, which simulates the highlands of its location. The texture, which is a revolutionary engineering technique, is based on the movement of water and generates a "diamond" pattern, transmitting luxury.

Our design accomplishes our main objectives: it shows up to date for a Premium water, reflects the luxury concept for its texture while the bottle's profile respects tradition.

Client
Aguas Minerales de Caldes de Boí, S. A.

Design Agency
Emmaolivèstudio (Spain)

Creative Director
Emma Olivè

Designer
Emma Olivè

Photography
Mauricio Fuertes

Technical Designer
Eva Fuertes

echo

Client
echo Beverages

Design Agency
Ferroconcrete (USA)

Creative Director
Yolanda Santosa

Designer
Yolanda Santosa, Sunjoo Park, Wendy Thai, Ann Kim

What makes echo stand out from other bottled waters is its simplicity; it is bottled water that minimizes its environmental impact. Clean, modern and responsible – the blue and green color palette evokes the purity of the water and the design of the packaging keeps waste to an absolute minimum. Even the quick-peel removable label was created to save time and effort; the labels were also printed in a carbon neutral, wind-powered facility As its tagline states, echo is "simple, local and responsibly packaged."

Sting Energy Drinks

Concept for a new line of energy drinks, Sting. The brief included coming up with a name for the drinks, logo design and producing labels for the cans. The target consumer is teenagers and young adults.

Client
Sting

Design Agency
Steph Baxter (New Zealand)

Designer
Steph Baxter

Photography
Tony Brownjohn

Juice "Juicy"

The "Juicy" brand is the 1st brand on the juice market of Kazakhstan. Under a complex service of the "Raimbek bottlers" brand portfolio, KIAN brand agency realized a re-positioning, redesign of logotype and package of the "Juicy" trade mark. The base of a new brand positioning became an idea of the "Juicy" TM fabulosity. It accentuated a rich history and the highest quality of "Juicy" juice on the market. Visual conception espouses the idea of legendary brand. KIAN brand agency designed bright off-beat cultic package. "Juicy" is distinguished from archaic images of Competitors by interesting food-style and courageous color layout of brand.

Client
"Raimbek-group" company

Design Agency
KIAN brand agency (Russia)

Creative Director
Kirill Konstantinov

Design Director
Maria Sypko

Designer
Anastasya Peskova,
Evgeny Mogalev

Photography
Anastasya Peskova

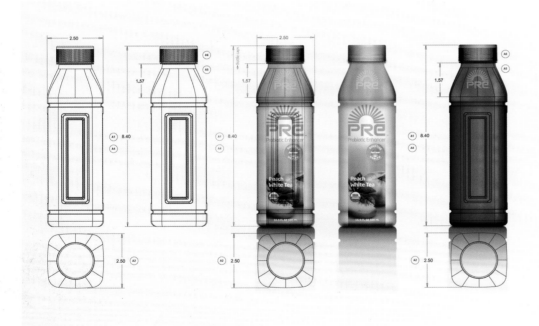

Pre-Branding and Packaging

Experience a new kind of beverage, PRE Probiotic Enhancer, an amazingly delicious fusion of organic juice enhanced with a proprietary prebiotic blend developed by Jarrow Formulas. PRE prebiotic beverage was developed to conveniently deliver you a healthy digestive boost any time of the day. PRE actively works with the body to promote healthy digestive balance by triggering the growth of beneficial bacteria called probiotics already presenting in your digestive tract. As the good bacteria increase so does resistance to harmful bacteria causing a boost in your immune system. PRE works with your digestive system helping to restore balance, creating a healthier and happier you. PRE. – Before Everything.

Client
Pre

Design Agency
TomTor Studio (USA)

Creative Director
Tom Tor

Designer/Photography
Tom Tor

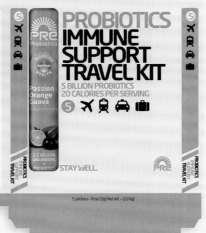

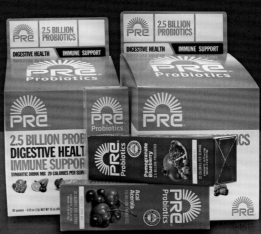

Acerola Logo and Wordmark

Due to an aging core demographic, loss of market share and weak sales, Drill Inc. partnered with Hoffman Creative on the exploration of a new identity to revive the once flourishing Acerola performance drink brand. A simplified icon reflecting the Acerola berry's structure was established for quick recognition in conjunction with an updated wordmark in a signature red and white palette. Following the packaging relaunch and a successful ad campaign, Acerola saw a 240% year-on-year increase in shelf space and huge spike in the drink's popularity amongst their target consumer.

Client
Drill Inc. for Nichirei Foods

Design Agency
Hoffman Creative (USA)

Creative Director
Morihiro Harano, Alex Freund

Art Director
Eric Hoffman for Hoffman Creative

Designer
Noriko Okamoto, Yasuo Ogusu,
Rie Kariya, Makoto Suzuki

Photography
Timothy Hogan

Copy Writers
Noriko Yamada, Hiroyo Kanehako,
Hidetoshi Kuranari

New palinka bottle

This bottle is a reinterpretation of the traditional bottle of Palinka, a hungarian fruit brandy.

Design Agency
kissmiklos (Hungary)

Creative Director
Miklós Kiss

Designer/Photography
Miklós Kiss

The Sopocani Juice

We designed a new brand mark and packaging, bottles for juices and jars for Serbian delectably products, made in the Monastery Sopocani. All products are organic and natural. The label is handwritten, implementing the spirit of the product.

Client
Monastery Sopocani

Design Agency
Peter Gregson Studio (Serbia)

Art director
Jovan Trkulja

Designer
Jovan Trkulja & Marijana Zaric

Honey Sample Set

For Beehive's varietal honey gift set, Hoffman Creative sealed 9 individual glass vials with synthetic stoppers and beeswax. The vials are presented in an e-flute carton with an illustrated raw brown paper insert and displayed in an American oak block.

Client
Beehive Beeproducts NYC

Design Agency
Hoffman Creative (USA)

Creative Director
Eric Hoffman

Photography
Timothy Hogan

Babees Honey

A simple and litlle idea of packing for honey. We tried to treat jar as a playground for a character design. Dark cap and stripes make the idea quite clear. Through this project we tried to encourage kids to reach for honey instead of refined sugar. Especially for them we create logo with a smiling bee's face. It's hand-calligraphed, custom-made, and it softens the overall simple, geometric look of the packaging.

Client
Ah&Oh Studio

Design Agency
Ah&Oh Studio (Poland)

Creative Director
Magdalena Kalek, Kamil Jerzykowski

Designer
Magdalena Kalek, Kamil Jerzykowski

The Europa Cafe

Client
Ken and Lidiya Dobosh,
Café owners

Design Agency
Camila Drozd, Freelance Studio (USA)

Designer/Photography
Camila Drozd

The Europa Cafe is a charming little place located on Main Street in Stroudsburg, which offers a wide range of delicacies from across Europe. Of my own accord, I redesigned the logo to better reflect the style and feel of the cafe and offered the design to the owners free-of-charge. Although they loved the logo, Ken and Lidiya decided not to take my offer due to the potential expense of changing everything. My offer is still on the table if ever they change their minds! The designs I created simulate what I hope the cafe sign and packaging will look like in the future.

True Rum | Promotional pack

Concept and design of the packaging to promote the seal of quality True Rum. The pack contains coins made of chocolate and Caribbean rum by the chocolate artist Ramon Morató.

Client
True Rum, Ramon Morató

Design Agency
Zoo Studio (Spain)

Creative Director
Gerard Calm

Designer
Xavier Castells

Photography
Ivan Raga

TheSimplyBar

Brand identity and package design for a line of all natural nutritional bars.

Client
Digestive Wellness

Design Agency
Christian Hanson (Canada)

Creative Director
Christian Hanson

Designer
Christian Hanson

Lola's Sprinkle Box

Client
Lola's Cupcakes

Design Agency
Campbell Hay (London, England)

Creative Director
Charlie Hay

Designer
Charlie Hay

Photography
Yann Binet

Lola's Kitchen bake yummy cupcakes and bring them round to your house. The obvious popularity of this idea meant Lola's needed a strong brand identity to help support their growth.

We coined the line 'Hand Crafted Cupcakes' to communicate the essence of what they do, and designed them an identity to illustrate the nature of their business across all media, from a sugar coated logo to pastel coloured stationery and packaging, and even – quite literally – the icing on the cake.

Bin Ablan | Mum

Design and packaging for a pastry company from Dubai.

Client
Bin Ablan

Design Agency
Zoo Studio (Spain)

Creative Director
Gerard Calm

Designer
Xavier Castells

Photography
Blai Pratdesaba

Sultry Sally

The brief was to design a range or unique chip packages for Sultry Sally, a new-to-world range of low fat, high flavour tasting potato chips.

Varga Girl style illustrations of the 1940's we created as the focus and core idea for the packs made them gave Sally immediate personality, attitude and standout from existing products on the market. The product price point was at a premium so the quality, attention to detail and classic style were required to reflect that. Sally is illustrated interacting with each flavour in a different way, this gives good flavour differentiation while Sally herself is a great link between all the packs in the range.

Title
Sultry Sally

Client
Potato Magic

Design Agency
The Creative Method (Australia)

Creative Director
Tony Ibbotson

Designer
Tony Ibbotson

Arctic Candy packaging

Sugarfree Arctic Candy are made of berries growing north of the Arctic Circle. Cooled by the arctic wind and warmed by the midnight sun, these berries ripen at a slow pace and grow to be sweeter than other berries. Extreme weather makes extreme berries. These are the extra sweet berries Eivind Glad use to make this candy. Neue redesigned the packaging which some of us started on in 2007.

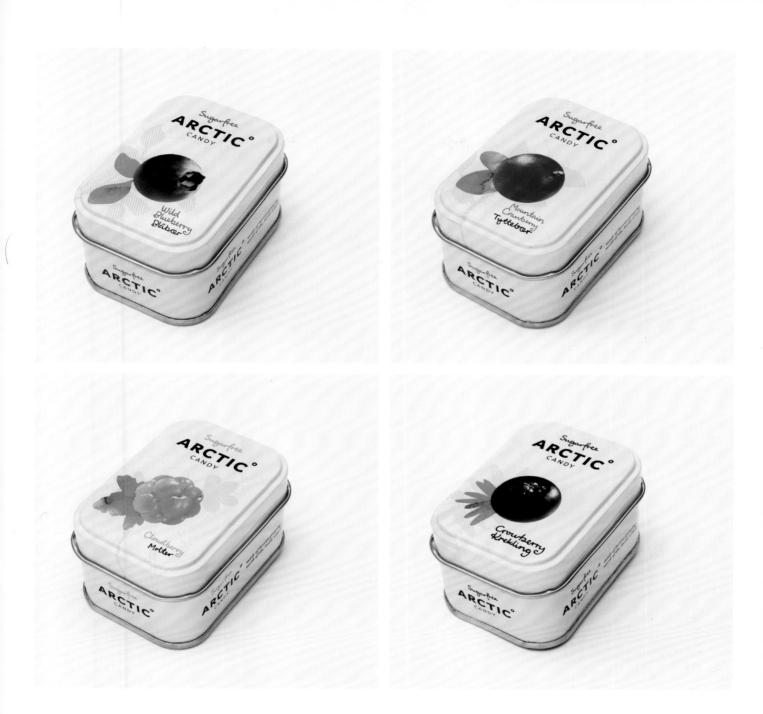

Client
Arctic Candy
Design Agency
Neue Design Studio (Norway)
Designer
Lars Havard Dahlstrom, Jostein Sandersen
Photography
on boxes: Lisa Westgaard, of boxes: Thomas Brun

Amazing Grass

Amazing Grass is a small San Francisco-based company whose mission is to help its customers live fuller, healthier lives through the miracle of wheatgrass. This hypothetical redesign includes a new brand identity, canisters for the powdered drink mixes, wrappers for the energy bars, and a promotional tote bag and bottle sleeve for healthy habits on the go.

Client
Katie Hatz

Design Agency
Katie Hatz (USA)

Design Director
Alice Drueding

Designer
Katie Hatz

Photography
Brandon Jones

Don't mess with my food

SUMM is a new brand of FairConnect that brings to market tasteful and fair food for the retail and out-of-home channels. In cooperation with BeyenMeyer and Conclusion Communication Architects, Reggs developed a brand that is reliable and honest about the source of the ingredients and the way it is cultivated. SUMM products are pure, very tasteful and offer a transparent chain perspective for consumers to contribute to a better world. The products contain biologically cultivated as well as fair-trade ingredients. Reggs also designed a line of differentiating retail packages. The brand has been exploited in a unique way: producers, marketers, consumers and creative agencies are working together to develop an innovative brand as a team effort.

Client
FairConnect

Design Agency
Reggs (The Netherlands)

Creative Director
Daan Huet

Design Director
Jop Timmers

Designer
Daan Huet

Photography
Ingemar Paalman

BONBON AU CHOCOLAT

BONBON AU CHOCOLAT is a packaging concept for "die Wäscherei"—a Lifestyle and funky furniture store in Europe. The aim was to create a visual concept that focuses on addiction of chocolate and transfers it into a modern, rock-n-roll kind of context. The protagonist looks like stylish junkies after the chocolate flush. Bonbon au chocolate – ambassador of love, pathfinder of lust. Bonbon au chocolate merges into us. Not any lover could ever treat us like this. It melts into us on a billow of bliss. There is no need to resist. Approving – Devoting – Sensing, Bonbon au chocolate and "Jet'adore" on the tip of the tongue. Life is too short for omitting a sin! Recommended by "Die Wäscherei".

Client
Die Wäscherei, das Möbelhaus - Hamburg

Design Agency
TIBOR+ (Germany)

Creative Director
Tibor Hegedues

Design Director
Tibor Hegedues

Designer
Anton Heinrich

Photography
Chon Choi / David Königsmann

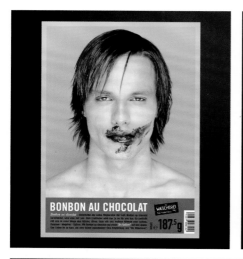

Client
"Trade Home Yarmarka" company

Design Agency
KIAN brand agency (Russia)

Creative Director
Kirill Konstantinov

Design Director
Maria Sypko

Designer
Evgeny Mogalev

Photography
Anastasya Peskova

"Yarmarka Café"

Under this trade mark "Café" Customer produces a very interesting and, as a tradition, absolutely new product for the Russian market. It is a unique collection of mix for delicious and interesting dishes such as polenta, paella, risotto, couscous, emmer, etc. It is offered to consumer not only as a fast food, but also as a product that helps to express their creative abilities and to realize their culinary talent giving scope for fantasy. It is as selected a conception of a comfortable café like a place where you can not only have a bite quickly and tastefully but also have a good time. Selected name – Café – perfectly opens the brand conception and creates a certain tint of preciosity and European style.

Haas Packaging

We first designed the mould for a chocolate bar in which no two pieces are alike, turning sharing into a game. Next, we designed a series of packaging for the diverse array of products included in the Thomas Haas chocolates line. The Percentage series of chocolate bars emphasizes the importance of cocoa content on the chocolate experience. A Tea series of chocolate bars focuses on various tea powders. For powdered hot chocolate, a can is made oval-shaped to maximize shelf presence. Finally, the Bites series and the Bark series are distinguished by their unique patterned windows which display Haas's chocolates.

Client
Thomas Haas

Design Agency
Bricault Design (Canada)

Creative Director
Marc Bricault

Photography
John Sinal

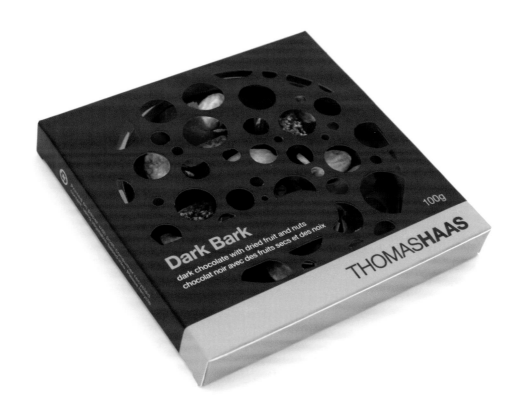

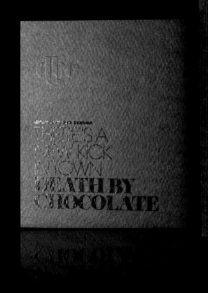

dbc – death by chocolate

Death By Chocolate is a corporate design and packaging concept for a fictional chocolate label. The aim was to create a unique packaging concept, which focuses on the myth of chocolate but transfers it into a modern, rock n roll kind of context.

Client
self initiated

Design Agency
dfact (Germany)

Creative Director
Denise Franke

Designer
Denise Franke

Photography
Denise Franke

MELT Chocolate

Melt, a gourmet chocolate shop, plays with the characteristics of melted chocolate with its logo and packaging in a fun and sophisticated way. The bars patterns look as if they were illustrated with warm, liquid chocolate while the tins have the look of being dipped. Even the truffle box has shiny chocolate finger prints as if it was just given to you by the chocolate maker. The packaging serves not to outdo the actual chocolate, but instead mimics the product serving to reinforce it.

Client
MELT

Design Agency
JJAAKK (USA)

Creative Director
Jesse Kirsch

Designer
Jesse Kirsch

GALERIE AU CHOCOLAT PACKAGING

Galerie au Chocolat has been one of our clients for the last four years and, in that time, we have designed a variety of packaging styles for them. They have a wide product range: some are available only in exclusive gourmet boutiques, while others are even sold at Costco. But Galerie au Chocolat believes in beautiful packaging and in an intelligent design at all times. From a classic line of fine black chocolates to a big pot of fair-trade hot chocolate, every product deserves the best, which we strive to provide.

Client
Galerie au chocolat

Design Agency
Paprika (Canada)

Creative Director
Louis Gagnon

Designer
René Clément

Theurel & Thomas

White is a central part of the design and it contrasts with the French macaroons colors. Details are the essential part of our work. We meticulously select each piece creating a balance with sophisticated specks that make the value of the brand and the exclusivity of the product outshine. In this project we worked in collaboration with Roberto Treviño & German Deheza.

Design Agency
Anagrama (Mexico)

Designer
Sebastian Padilla,
Miguel Angel Herrera,
Roberto Trevino,
German Deheza

Photographer
Caroga Foto

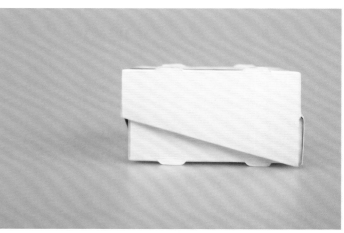

RESCHE RESI

This is a non-glued package for handmade zwieback. It is not only for protecting the product, but also a good way for presenting and serving.

Client
self initiated work

Design Agency
kopfloch.at (Austria)

Creative Director
Gerlinde Gruber

Designer
Gerlinde Gruber

Photography
Stephan Friesinger

64

Client
Rio Coffee

Design Agency
Voice (Australia)

Creative Director
Anthony De Leo, Scott Carslake

Design Director
Anthony De Leo, Scott Carslake, Shane Keane

Designer
Shane Keane

Rio Coffee has been hand-roasted Australia's favourite Coffees since 1964, with 14 varieties in their portfolio. 64 presents a premium coffee product that reinforces Rio Coffee's longevity in the coffee industry and evokes a sense of nostalgia by looking back at the birth of the brand.

The aim was to create a premium, contemporary solution whilst maintaining a style reminiscent of the 60s. It uses simple geometric shapes and colours to represent the rising aroma and steam from a cup of coffee.

HAPPY AND SWEET 2009

This is a corporate gift, designed to wish clients a Happy and Sweet 2009. It is a box with a set of 12 chocolate bars, which represent the 12 months of a year, each with a different illustration that incorporates an element the month (like a rose for Valentines Day in February). It works as a calendar… and the chocolate has to be eaten at the end of each month.

Client
Kanella – Self Promotion

Design Agency
Kanella (Greece)

Creative Director
Kanella Arapoglou

Designer
Kanella Arapoglou

ICA Candy

ICA Candy Candy not only tastes good, it has to be fun too. On a crowded store shelf ICA´s own brand candy vies for space with cartoon animals and fairytale characters.

Silver made the new ICA designs stick out from the competition by bringing alive the contents of the package.

 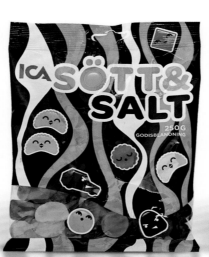 

Client
ICA Sweden AB

Design Agency
Designkontoret Silver KB (Sweden)

Creative Director
Ulf Berlin

Design Director
Cajsa Bratt

Client
PROBAR

Design Agency
Moxie Sozo (USA)

Creative Director
Leif Steiner

Design Director
Leif Steiner

Designer
Nate Dyer

Photography
n/a

Fruition Packaging

Probar is a high performance energy bar made from whole, raw foods. Each flavor is a different mix of fruits, nuts, berries, and seeds. Unlike most energy bars, Probar's ingredients are natural and recognizable.

Moxie Sozo was hired to redesign Probar's entire product line. After the new launch, sales increased dramatically, market share improved, and the company picked up numerous new retail accounts.

Client
Biscuiteers

Design Agency
Big Fish ® (UK)

Creative Director
Perry Haydn Taylor

Design Director
Victoria Sawdon

Designer
Victoria Sawdon

Photography
Big Fish ®

BISCUITEERS CONRAN TIN

A brand new concept for the UK: themed biscuits, beautifully iced and gorgeously presented. The strap line? "Why send flowers when you can send biscuits instead?" A product with this much personality needed branding and packing to match. So, Big Fish developed a charming, quirky illustrative style, which informed everything from the logo to the tin labels. The collection pictured was an exclusive design for The Conran Shop.

Fibar

Package and identity design for a line of all natural fiber bars.

Client
Wellness Foods

Design Agency
Christian Hanson (Canada)

Creative Director
Christian Hanson

Designer
Christian Hanson

Dent oi... så store

Dent oi... så store candy is a unique new lozenge with an exciting sweet taste. It is surprisingly big – hence the name (oi... så store = oops... so big). The design emphasizes the product´s distinctiveness in a simple but engaging way. The packaging informs the consumer that this is an intriguing, fresh and special lozenge in keeping with Dent´s personality. Dent oi comes in two different varieties - fuzz and sour.

Client
Brynild Gruppen

Design Agency
Strømme Throndsen Design (Norway)

Creative Director
Morten Throndsen

Design Director
n/a

Designer
Linda Gundersen

Photography
n/a

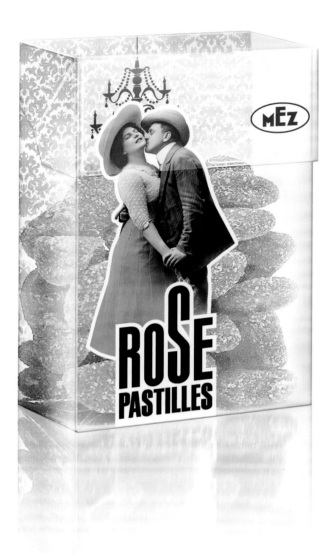 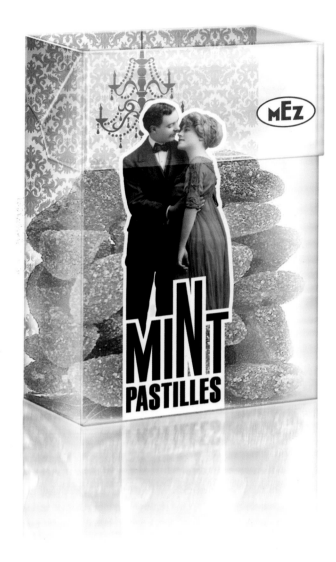

Mez Candies Box

Client
Lavdas Sa

Design Agency
Mouse Graphics (Greece)

Creative Director
Gregory Tsaknakis

Design Director
Gina Zafiraki

Designer
Vassiliki Argyropoulou

Illustrator
Ioanna Papaioannou

Redefining the image and proposition of a very traditional Greek soft candy, originally named Kiss or Date, the product captures both past and future generations with its revived tradition. The pack evokes a feel of nostalgia; viewers are drawn to it through a retro romantic set-up, almost subliminally aware of the brand's past.

Tesco Cookie & Doughnut Bags

Following declining sales, Tesco briefed us to inject some fun into their own label cookie and doughnut packaging to increase shelf standout and to generate more interest in the category at large (the products are merchandised at the end of aisles). The prerequisite was a window in the packaging to physically show the freshness of product due to limited shelf life.

We created a fiendishly simple solution that made relevant use of the required transparent window by showing the products in the stomachs of a fun doughnut and cookie monster. The term cookie/doughnut monster is UK/US slang for someone who eats a lot of cookies or doughnuts.

Client
Tesco

Design Agency
Taxi Studio (UK)

Creative Director
Spencer Buck

Designer
Roger Whipp

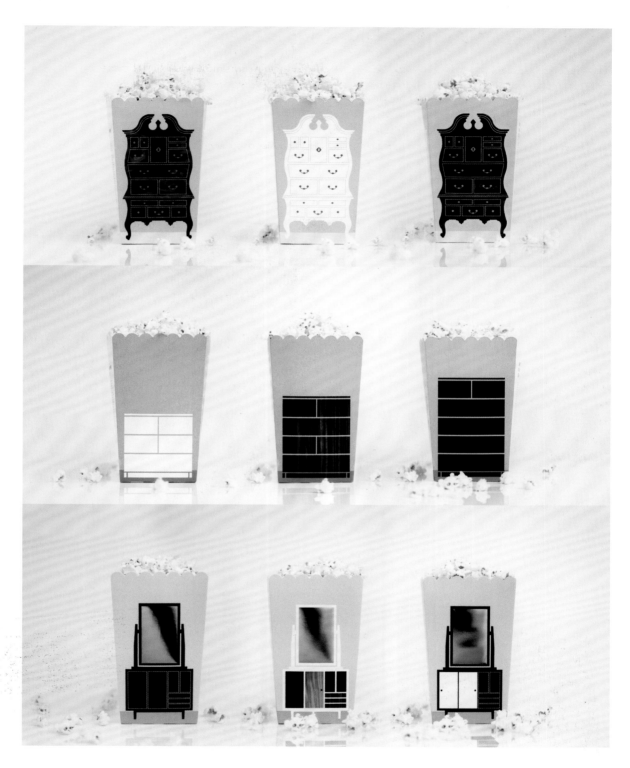

Client
Elanders/Bureau of the year

Design Agency
Milk (Sweden)

Designer
Martin Ohlsson, Mikael Selin

Photography
Martin Ohlsson

Bureau of The Year

Packaging design for the annual Bureau of The Year gala in Sweden. A gala where advertising, design, PR and media agencies are rewarded based on client ratings.

Hand Made Biscuits

We want to give to a pure and traditional biscuit's a totally different image, modern and clean but in the same time strong, for that reason we work with bold colors. The tin is to keep the product in perfect conditions. This box wan a One Show Design merit award and a Sena da Silva Award 2009 from the Portuguese Design Center.

Client
Boa Boca Gourmet

Design Agency
António João Policarpo
/ Policarpo Design (Portugal)

Design Director
António João Policarpo

Designer/Photography
António João Policarpo

  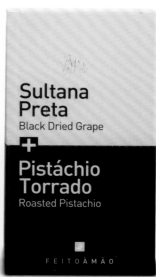 

Client
Boa Boca Gourmet

Design Agency
António João Policarpo
/ Policarpo Design (Portugal)

Design Director
António João Policarpo

Designer/Photography
António João Policarpo

Dried Fruits

We have conceived one original package to split in 2 through a perforated line. After opening it, you will find inside 3 different dried fruits (sweet+salted) that might be eaten at the same time, enjoying the bittersweet sensation as if it were 2 small bowls of aperitifs. This box wan a Red Dot Award 2009, Communications Arts – Award of Excellence 2009 and Sena da Silva Award 2009, from the Portuguese Design Center.

KAFFEEWEIZEN

KAFFEEWEIZEN is a hand filled family business product. The design of the glass bottle is exclusive and reduced to the essential elements. Every year just a limited amount of these special coffee & corn liquor blend is available.

0,5 l TRADITION MIT GESCHMACK 30% vol
HERGESTELLT IN **NEUENRADE** GESTALTET VON **NALINDESIGN.COM**

Client
KAFFEEWEIZEN

Design Agency
Andre Weier / NALINDESIGN (Germany)

Creative Director
Andre Weier

Designer
Andre Weier

Carapina

Carapina is the name of the homemade high quality Italian gelato brand. The history of the name is connected with the storic gelato sellers, witch were travelling from place to place with their little gelato trolley. Carapina gelato design makes you think about old gelato sellers, tradicional recipies and fresh ingredients, but at the same time is one of the most experimental and strategic gelato brands in Italy.

Client
Carapina

Design Agency
Doni & Associati (Italy)

Creative Director
Simonetta Doni

Designers
Francesco Graziani

Photography
Doni & Associati

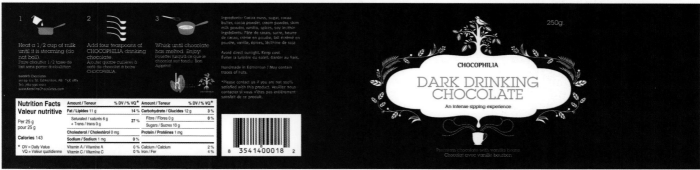

Chocophilia Drinking Cocoa

Client
Kerstin's Chocolates

Design Agency
Vanguard Works (Canada)

Creative Director
Keith-yin Sun & Judi Chan

Designer
Keith-yin Sun & Judi Chan

Illustrator
Keith-yin Sun & Judi Chan

Chocophilia is an artisan chocolate brand focuses on high quality gourmet chocolate products with unique recipes and natural ingredients. We custom-created this decorative motif with elements based on a cocoa plant, while in the middle is a transformation from the tree's root to drippings of steaming hot Drinking Cocoa into a cup. The idea is to blend the delicacy of the quality of the product and the natural aspect of the origin of the cocoa with antique inspired decorative elements and a earthy palette on a stock with rich natural texture.

Chocophilia Cocoa Nib Caviar

As a sister product to the Chocophilia Drinking Chocolate line, we have based the design largely on the Drinking Chocolate. Due to the smaller surface area, we have reduced the decorative motif to only simple shapes, still showcasing the Chocophilia Tree, surrounding now are stylized "pearls" symbolising the Cocoa Nib Caviar, with a victorian style caviar serving dish at the bottom, completing the cycle "from the cocoa tree to the product".

Client
Kerstin's Chocolates

Design Agency
Vanguard Works (Canada)

Creative Director
Keith-yin Sun & Judi Chan

Designer/Illustrators
Keith-yin Sun & Judi Chan

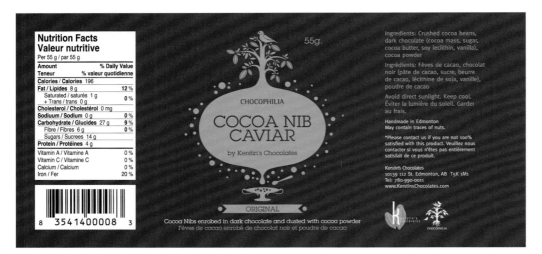

Client
Brynild Gruppen

Design Agency
Strømme Throndsen Design
(Norway)

Creative Director
Morten Throndsen

Design Director
n/a

Designer
Linda Gundersen

Photography
Strømme Throndsen Design

Minde Sjokolade

Minde sjokolade has relaunched its chocolate bags with great success. Even well known and trusted brands need to be updated to keep themselves fresh and relevant. The new environmentally friendly bags developed by Strømme Throndsen Design have visual variety reflecting the content and individual personality. Colours and details contribute to the range's high level of shelf presence. The content in each of the chocolate bags is communicated through the use of different colours and design elements. The common pack feature is a detailed product presentation of the individual chocolates making it easy for everyone to recognize and enjoy their favourite.

Client
Deseo Srl

Design Agency
Doni & Associati (Italy)

Creative Director
Simonetta Doni

Designers
Simonetta Doni, Giuliano Fenn

Photography
Doni & Associati

Deseo

Deseo biscuit line contains:
- Breakfast biscuits (gold packets)
- Cantucci and butter biscuits in PVC packets
- Cantucci and butter biscuits in boxes (for luxury food shops)
- Gift boxes (steel box)
- Olive Oil biscuits (boxes with "oil drop seal")
- Aperitif salted biscuits (black packets)

We have characterized every line of biscuits by the different package material and/or the background colour.

The cantucci and butter biscuits boxes are designed for shops specialized in luxury food products. The different flavour of the product is evidenced by the sample picture on the top of the box.

Gift steel boxes contain several kinds of Deseo butter biscuits: Pure butter biscuits, Chocolate chips biscuits, Biscuits with coffee, Biscuits with cinnamon. This confection can be a perfect gift. Once you finish your biscuits, you can take off the ring-shaped label and use the box however you wish.

Our following luxury line are sweet biscuits with extra virgin olive oil. To emphasize that the biscuits contain the tuscany extra virgin olive oil, we have designed a 3D plastic oil drop-shaped seal.

The product line of Salted biscuits with sesame seeds and paprika is ideal for any kind of aperitif. We have chosen black for package color to underline the elegance of the product and to distinguish these spicy biscuits from the sweet ones.

Jan Robben

Jan Robben is a strawberry grower with a passion for his product, and a dedication to producing fruit of a higher standard than your 'average' strawberry. In 1998, his company was the first in the Netherlands to receive the Milieukeur certificate. We designed a packaging that presented his strawberries in a way similar to a box of chocolates – as an exclusive, luxurious treat rather than a commodity. The packaging has a natural, intimate look-and-feel, featuring birds, Robbens' butterfly logo and his signature, reflecting his personal devotion and care for the environment. This enabled Jan Robben to change his market strategy and achieve significant business improvement.

The Milieukeur certificate is awarded for special efforts in the field of natural, environmentally friendly production processes. Milieukeur products are very similar to organic products, but a limited amount of chemical crop-protection agents and artificial fertilisers are permitted in their production.

Client
Jan Robben

Design Agency
Reggs (The Netherlands)

Creative Director
Daan Huet

Design Director
Jop Timmers

Designer
Daan Huet

Photography
Ingemar Paalman

Client
Raw Health

Design Agency
Pearlfisher (UK)

Creative Director
Natalie Chung

Design Director
Sarah Pidgeon

Designer
Vicki Willatts

Raw Health – Vibrant Living

Pearlfisher was tasked with creating an identity and packaging solution that redefined the category: taking the raw food movement beyond its current niche and making it universally understood and accepted. The essence of RAW is taking food at its most primary state, that is, uncooked, basic, untouched and unprocessed — keeping in all the goodness and nutrients, as opposed to cooking or heating them out.

The contemporary packaging reflects the edginess of the brand and literally makes RAW raw. It expresses a definitive look and feel of 'vibrant living' through a palate of rich colours and bold expressive typography. The tone of voice captures the energy, vitality and 'packed-with-as-much-goodness as there can possibly be in one product' sentiment. It is as real as it is raw.

The Wanted Snacks

The objective for this project was to create an eye catching package for the most desired (wanted) products, that's fun and communicative. The warrant poster is the idea that connects packaging and communications as common denominator, conceptually and visually. Each product has a warrant for a wanted person on the front. For instance cashew nut is an Indian. In Serbian, a cashew nut is translated as Indian nut, which explains the turban.

Client
Aroma Food

Design Agency
Peter Gregson Studio (Serbia)

Art Director
Jovan Trkulja

Designer
Jovan Trkulja & Marijana Zaric

Prodiet

Prodiet is a diet-food brand from Slovenia. It is positioned as a healthy product with a great taste.

Client
Difar Ltd.

Design Agency
Studio 360 (Slovenia)

Creative Director
Vladan Srdic

Designer
Vladan Srdic

Photography
Luka Kase

Client
Holli Mølle

Design Agency
Strømme Throndsen Design
(Norway)

Creative Director
Eia Grødal

Design Director
n/a

Designer
Linda Gundersen

Photography
n/a

Holli Mølle

Holli Mølle flour originates from the Holli Mølle organic mill in Eastern Norway. It specializes in the use of ancient and nutritious grain types in its production of flour. Strømme Throndsen Design has created a strong identity and packaging design, based on the mill's core values of tradition, organic production, authenticity and environmental awareness. The packaging stands out on the shelves and is functional, flexible and efficient in production. Bright label colours work as a differentiator among the 6 variants. The design communicates well with the target group, giving them a feeling that the flour really is 'ground with love', as stated in the personal message from the owner, Trygve Nesje.

"Yarmarka Platinum"

Under "Yarmarka Platinum" trade mark, the client launched a new series of premium grain traditionally used in national cuisines all over the world. A complex development of the brand was realized by KIAN brand agency. Developed visual conception creates a unique, striking personality of the products and corporate identity. The products' attractive textured packages bear unusual form, color, and graphic symbols.

Client
"Trade Home Yarmarka" company

Design Agency
KIAN brand agency (Russia)

Creative Director
Kirill Konstantinov

Design Director
Maria Sypko

Designer
Evgeny Mogalev

Photography
Anastasya Peskova

Other
Gold Pentawards 2009

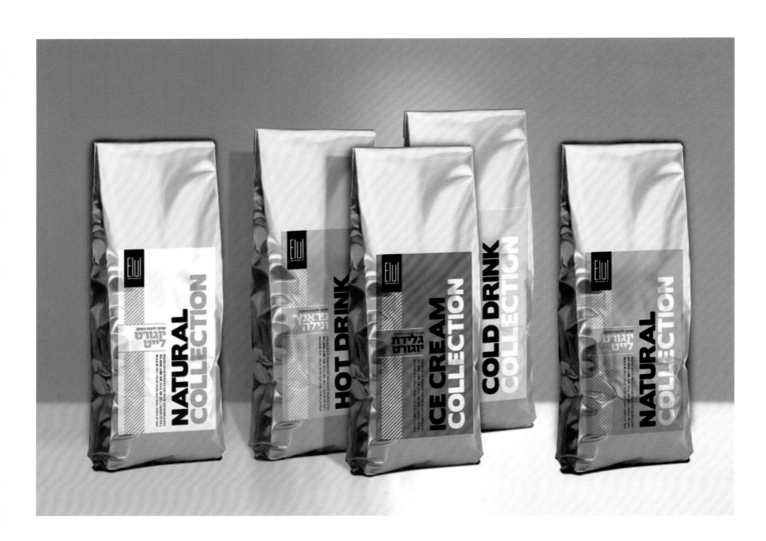

Graphic Design & Branding

Branding for a food industry company.

Client
elul

Design Agency
Artilerya (Israel)

Designer
shlomi betito

Client
Floridabel

Design Agency
Peter Gregson Studio (Serbia)

Art Director
Jovan Trkulja

Designer
Marijana Zaric

Fun&Fit Cruchy Muesly

New packaging design for "Fun&Fit" Crunchy Muesli for Serbian company 'FLORIDAbel'.

There are three different tastes: Chocolate, Fruit and Honey, and Crunchy Muesli.

Fusión Mate de Coca

Package and identity design for a herbal energy tea for the South American market.

Client
Fusión Mate De Coca

Design Agency
Christian Hanson (Canada)

Creative Director
Christian Hanson

Designer
Christian Hanson

CLIPPER

Client
Clipper

Design Agency
Big Fish ® (UK)

Creative Director
Perry Haydn Taylor

Design Director
Victoria Sawdon

Designer
Victoria Sawdon

Photography
Big Fish ®

Re-brand and package Clipper's range to truly reflect the company's Fairtrade heritage and natural approach to creating utterly delicious teas. The aim was to create something people would be proud to have on display, rather than hide away in their kitchen cupboard.

Premier Garden Tea Collection

Developing Premier Garden collection, Kollor's assignment was to develop a packaging that would radiate classical elegance and high quality, combined with a modern approach. Kollor has also produced brochures and marketing material for the shops, and has created TeaButlers website too.

Client
Teabutler Scandinavia

Design Agency
Kollor Design Agency (Sweden)

Creative Director
Erik Tencer

Design Director
Erik Tencer, Åse Ekström

Designer
Åse Ekström, Håkan Persson

Photography
Petter Ericsson, Freddy Billqvist

Sugar Sticks

This work is a conscious attempt to literally interpret the pack content: 'One spoon of sugar' – visually and verbally.
Straightforward and simple, the pack design obviously introduces a more delicate sugar stick.

Client
Sugarillos Sa

Design Agency
Mouse Graphics

Creative Director
Gregory Tsaknakis

Design Director
Gregory Tsaknakis

Designer
Maria Karagianni

Other
Illustrator Ioanna Papaioannou

Jaipur Avenue Chai Tea Packaging

The faraway Indian city of Jaipur, famous for its royal palaces, colorful culture and romantic charm epitomizes the ancient chai tradition at its best. Jaipur Avenue™ is your instant passage to this magical land through an enthralling all-natural chai experience you can enjoy anywhere, anytime. As diverse as the colors of India, so are the flavors of Jaipur Avenue chai: Ginger, Cardamom, Masala, Vanilla and Saffron.

Client
Jaipur Avenue

Design Agency
Turnstyle (USA)

Design Director
Ben Graham

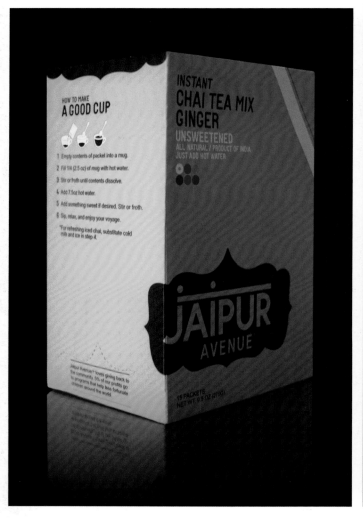

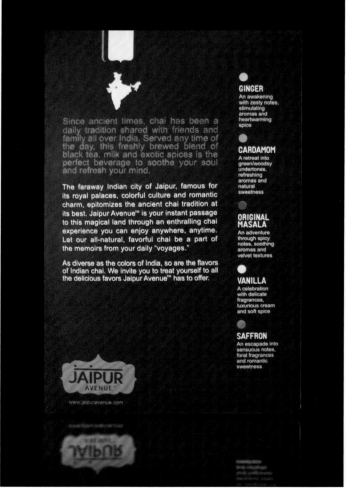

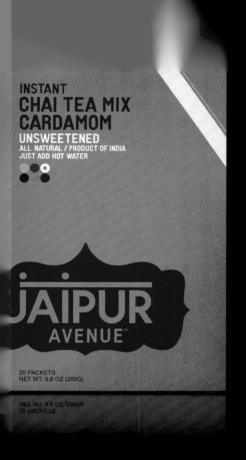

What On Earth organic cake packaging range

We designed a new packaging concept for the organic food producer "What on Earth". The new packaging uses linocut illustrations that can be arranged in different ways. It is important not to use computer-generated but hand-made images that relates to the hand-made and organic aspect of the product. The cake boxes are the first new design in a range; other products packaged in boxes, tins and wrappers will follow later. The principle of using lino cuts illustrations from a now established database of images on a plain background allows designs of different types of packaging easily.

Client
What On Earth

Design Agency
Mind Design

Designer
Holger Jacobs, Craig Cinnamon

Mekong Red Dragon Rice

Client
Xanh Vietnam

Design Agency
Scott Lambert (Vietnam)

Designer
Scott Lambert

Photography
Kevin German

Illustrator
Andrew Denholm

Mekong Red Dragon Rice

The rice is Eco friendly and Fair Trade. Furthermore husking, sifting and packaging are done by local villagers creating additional employment. It is also healthier Rice; during the husking process, the bran layer is only partly removed, as it is within the bran layer that the majority of beneficial nutrients are stored.

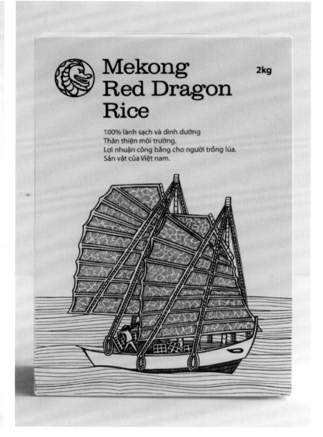

Koberg Vilt

Koberg Vilt is a traditional Swedish food with a twist. Wild boar salami with chili and deer sausages with Bourgogne wine are just two of the many products made from game living on the Koberg estates. The vivid hunting colours and classic oak leaf patterns make the Koberg product line eye-catching.

Client
Koberg Vilt

Design Agency
Designkontoret Silver KB (Sweden)

Creative Director
Ulf Berlin

Design Director
Cajsa Bratt

ICA Chrunchy

You shouldn't play with food...or maybe you should! Playfulness and cleverness turned out to be a great concept for the Swedish grocery chain ICA's cereal product Crunchy.
The playful photos are displayed all around the packaging and the products of the range make up a long breakfast table when put together, no matter in what order. The photographer get grey hairs, ICA get a very attractive cereal shelf where their private label put the A-brand of the category in the shade. And the consumer get a breakfast with a smile.

Client
ICA AB

Design Agency
Designkontoret Silver KB (Sweden)

Creative Director
Ulf Berlin

Design Director
Steven Webb

Photography
Roland Persson

Client
Vitacress

Design Agency
Big Fish ® (UK)

Creative Director
Perry Haydn Taylor

Design Director
Victoria Sawdon

Designer
Ruth Brooker

Photography
Big Fish ®

STEVE'S LEAVES

Create a striking new salad brand to stand out from 'the hedge': the mass of green leaves in the chiller cabinet. The scheme uses bold colours and unique hand-written text to deliver the key messages of pioneering, tasty, nutritious, and environmentally conscious.

Client
Norgesgruppen

Design Agency
Strømme Throndsen Design (Norway)

Creative Director
Morten Throndsen

Design Director
n/a

Designer
Linda Gundersen

Photography
Lisa Westgaard (Tinagent)

Jacobs Utvalgte

Jacobs Utvalgte (Jacob's selected) is a new range of premium, in-house brand products, developed by Strømme Throndsen Design, in cooperation with the largest retail grocery group in Norway, with a 40% market share—NorgesGruppen. The design communicates an immediate feeling of quality with its simplistic but striking combination of black surfaces and tempting food pictures. The guarantee symbol and information ribbon add credibility and trust to the brand, and help differentiate between each product consumer benefits. The packaging is easily recognizable across a range of product categories and has contributed to the huge success of Jacobs Utvalgte for the distributors Meny and Ultra of NorgesGruppen.

Jordans

Pearlfisher have created a natural world of Jordans, starting at the mill and continuing 360˚ around the pack with illustrations that reflect the personality of each flavour variant. The bold colours are crammed full of taste and are offset by the white sky which allows the Jordans logo more prominence than ever before.

Client
Jordan's Cereal

Design Agency
Pearlfisher (UK)

Designer
Will Gladden

Client
Nestlé Hungary Ltd. - Nesté Hungária Kft.

Design Agency
Café Design (Hungary)

Creative Director
Mr. Attila Simon

Design Director
Mr. Tamás Veress

Maggi Forró Bögre – Maggi Hot Mug

Our client is the world's leading nutrition, health and wellness company. Our task was to help the successful introduction of a new product family under the already existing Maggi brand by creating an attractive packaging.

Though Maggi already had international standards concerning design, the designer got relative freedom to create a 'Hot Mug' ('Forró Bögre') logo and illustrations.

It is really challenging to design the biggest possible logo and taste-illustrations for such a small surface. The obligatory yellow brand colour of Maggi is resembled on the mug itself thus giving space and freedom for new solutions in figures and colours.

SPICE BOX

Mini wooden box for spices.
Stackable and easy storage.
Dimensions: 50 x 50 x 50 mm.

Design Agency
Baita Design Studio (Brazil)

Creative Director
Helena Baita Bueno

Design Director
Heinz Muller

Designer
Helena Baita Bueno

Photography/Renders
Heinz Muller

1854 herbs and spices

New packaging for 1854 herbs & spices 40cc tins. The new label, developed by Estudio Clara Ezcurra with shiny black background contrasting with bright and metallic colors, gives the product a high quality finish and an innovative look. The design combines very simple non conventional associations for each spice with a silhouette made to identify every product. The main idea was to translate some humor signs of universal culture which addressed a diversity of locations, cuisines and regions: the Andes, the Pampa, southern and northern Argentina and other places and cultures from which the goods are original.

Client
Prosabores

Design Agency
Estudio Clara Ezcurra (Argentina)

Creative Director
Clara Ezcurra

Design Director
María Olascoaga

Designer
María Olascoaga

Photography
Facundo Basavilbaso

Glorious! Soup

ilovedust worked on GLORIOUS! Soup rebrand with TSC Foods and Lambie-Naim recently and were very proud of the results! An iconic illustration style was created for each letter of the alphabet, representing the authentic global influences and flavours of each product. For example, within the soup range, T is for 'Toulouse Sausage and Bean' and M stands for 'Malaysian Chicken'.

Mark Graham, Creative Director at ilovedust: "We had a lot of fun with the new GLORIOUS! branding with so many interesting ingredients to work with and global locations to take inspiration from, we were spoilt for choice on what we could do graphically."

Client
TSC Foods and Lambie-Naim

Design Agency
ilovedust (UK)

Creative Director
Mark Graham

Design Director
Johnny Winslade

Designer
ilovedust

Photography
n/a

Egizia

Starting from the idea of a pleasent and aesthetic solution to the presence of wine on the table beyond the logic of formal degustation, Egizia has decided to challenge the world tableware. With its acknowledged daring and determination, it has extended its research on objects to all those components, which did not have, until now, a common connotation, but rather presented themselves as dispersive and non-recognisable.

Client
Egizia Srl

Design Agency
Doni & Associati (Italy)

Creative Director
Simonetta Doni

Designers
Simonetta Doni

Photography
Doni & Associati

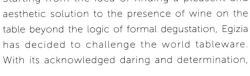

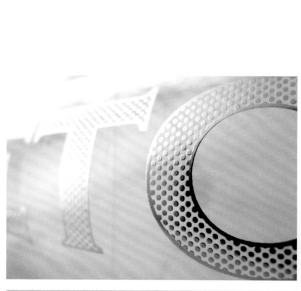
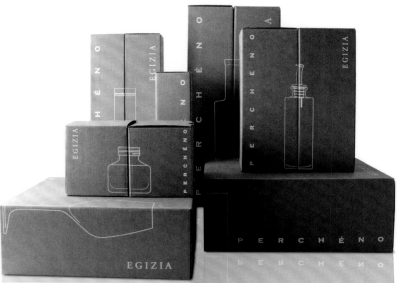

1854 Lemon Powder

Lemon powder packaging for 1854 herbs & spices. Our studio has designed the Lemon powder packaging for 1854's new product.
We were asked to make 3 presentations: Lemon powder, Lemon powder and pepper, Lemon powder and salt.

As the product is new in the market we have developed an eye-catching specially illustrated display for each item, changing only the color schemes according to the contents.

Client
Prosabores

Design Agency
Estudio Clara Ezcurra (Argentina)

Creative Director
Clara Ezcurra

Design Director
María Olascoaga

Designer
María Olascoaga

Photography
Andrea Saslavsky

Salsa Packaging

There are a wide variety of salsas in the marketplace, with offerings from small start-ups and international corporations alike vying for consumer dollars. Moxie Sozo wanted to create salsa packaging for Fruta Del Diablo that would distinguish it from everything else on the shelf and establish credibility for an unknown brand. By using hand-drawn illustrations inspired by the woodcuts of Mexican artist Jose Guadalupe Posada, we were able to lend authenticity to the salsa while reinforcing the product's heritage in traditional Mexican cuisine.

Client
Fruta Del Diablo

Design Agency
Moxie Sozo (USA)

Creative Director
Leif Steiner

Design Director
Leif Steiner

Designer
Nate Dyer

Photography
n/a

Salsa Packaging

LEAP Organics is a bath and body products company out of Boston, Massachusetts. Moxie Sozo began working with founder, Luke Penney, when the company was in the initial concept stages. Faced with a highly competitive landscape, the agency identified four goals for developing the brand: 1) Be bold. 2) Be different. 3) Take risks. 4) Get noticed. Moxie Sozo worked with LEAP Organics to develop a personality that was appropriate, yet highly differentiated from other products in the category. Keeping within the spirit of the brand, the LEAP Organics' soap packaging was illustrated entirely by hand without the use of a computer. Within weeks of launching the initial line of soaps, LEAP was picked up for distribution within Whole Foods Mid-Atlantic and other key retailers. In the coming months, Moxie Sozo and LEAP will be introducing a line of organic skin care products. The agency also handles all of LEAP's website, social media and public relations needs.

Client
LEAP Organics

Design Agency
Moxie Sozo (USA)

Creative Director
Leif Steiner

Design Director
Charles Bloom

Designer
Charles Bloom

Photography
n/a

Kavli Mayonnaise

Strømme Throndsen Design collaborated with Edge Consultants and Kavli on concept and packaging development, resulting in a highly innovative product with a patented adjustable closure. In just 6 monts Kavli Mayonnaise has reached a 12% market share by value- giving the market leader, Mills, a tough time.

Client
Kavli

Design Agency
Strømme Throndsen Design (Norway)

Creative Director
Morten Throndsen

Design Director
n/a

Designer
Nina Kristensen

Photography
Lisa Westgaard- Tinagent

Matstreif

Client
Innovasjon Norge

Design Agency
Strømme Throndsen Design (Norway)

Creative Director
Morten Throndsen

Design Director
n/a

Designer
Eia Grødal

Photography
n/a

Matstreif is the largest food festival in Norway taking place in September each year in the main street of Karl Johan in Oslo. Norwegian food, local food production and the joy of good food are the main themes of the festival arranged by Innovation Norway.

Every Day Pasta Sauce

"Every Day" - Svaki Dan is brand name for tomato based, handmade pasta sauces. Client is Danubius. There are four delicious different tastes.

Client
Danubius

Design Agency
Peter Gregson Studio (Serbia)

Art Director
Marijana Zaric

Designer/Ilustrator
Marijana Zaric

Fjordland Ready-to-eat meals

The real taste of a good meal from Fjordland is well known to Norwegian consumers. The new "4 minutes" range makes it even easier to enjoy a great Fjordland meal by the easy-to-heat packaging.
Strømme Throndsen Design has created a simple and down-to-earth design concept focusing directly on the Fjordland competive advantage of great taste. The food and the tray itself are the heroes of the design, easily accessible to consumers of all ages.

Client
Fjordland

Design Agency
Strømme Throndsen Design (Norway)

Creative Director
Morten Throndsen

Design Director
n/a

Designer
Jarle Paulsen

Photography
Lisa Westgaard (Tinagent)

AU Olive Oil

Olive oil is well known as liquid gold, being Au (Latin Aurum) the chemical symbol for Gold.
Objective: To seize that idea and create a unique identity. The golden brick recalls the purity of its contents: 100% Extra Virgin oil.

Client
Aceites Únicos

Design Agency
Wesemua (Spain)

Creative Director
Wesemua

Photography
Wesemua

EL VERD DEL POAIG

By pulling the cardboard petals off this exclusive packaging, we discover the slender figure and soul of El Verd del Poaig.
Recycling and luxury materials undertake a joint venture to reveal this unique olive oil with aromatic notes of freshly cut grass, walnuts and almonds.
Floral, intense, slightly bitter and spicy, sweet, fluid, just delicious... A classy oil in an ecofriendly packaging.

Client
El Poaig

Design Agency
Culdesac (Spain)

Creative Director
Culdesac

Designer/Photography
Culdesac

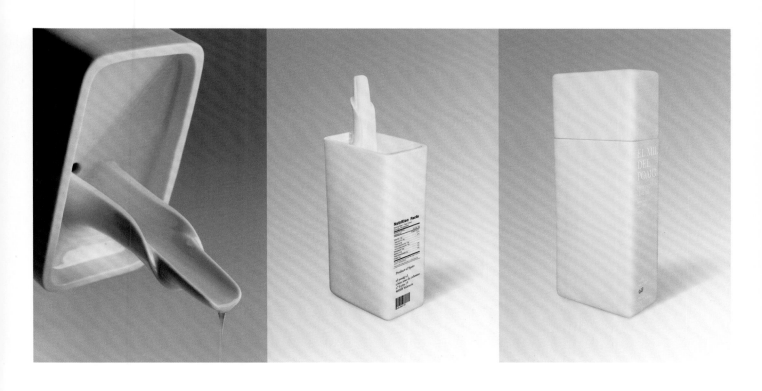
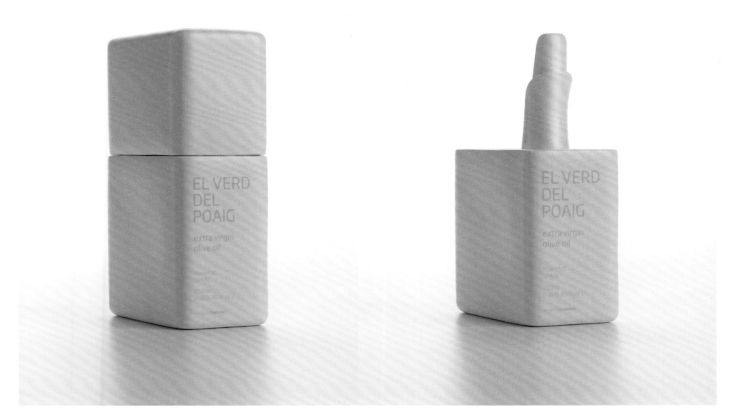

Client
Tenuta del Buonamico

Design Agency
Doni & Associati (Italy)

Creative Director
Simonetta Doni

Designer
Simonetta Doni

Photography
Doni & Associati

Oro di Re

We wanted to create an image with a great impact, the same image for supermarkets and for restaurant distribution. Three different destinations: D.O.P. Oil, Biologic Oil and 100% Italian Oil. "Oro di Re" means The Gold of the King, which is represented by the stilized gold crown used in the packaging.

Iliada Premium Olive Oil

Based on a brief that essentially could be distilled in two key words – Premium Quality & Visual differentiation – this olive oil packaging targets an international high-end audience. In an otherwise crowded shelf space where all fight to convince by visually stating the obvious, this packaging is deliberately moving away from traditional symbols of olive oil quality or clichés of provenance. This pack design targets the mind; aesthetics is the real reason to buy. Almost as an afterthought, a very realistic-looking drop of oil on the tin is what keeps it grounded into the food section of a super market. If ever there was a slogan attached to it, then this would read 'simply oil'.

Client
Agrovim Sa

Design Agency
Mouse Graphics (Greece)

Creative Director
Gregory Tsaknakis

Design Director
Vassiliki Argyropoulou

Other
Illustrator Ioanna Papaioannou

Quickoven identity and packaging design

Quickoven is a vending machine company that serves hot food in less than 90 seconds, around the clock. Since there is no human encounter involved, we decide to personalize the experience of ordering and receiving your meal. We did so by giving the machine and packaging a voice of its own, adding a bit of charm and humour to the process.

Client
Quickoven

Design Agency
Milk (Sweden)

Designer
Martin Ohlsson

Photography
Martin Ohlsson

Copywriter
Niklas Moberg

Our farm dairy

The small factory makes dairy production of the highest quality—without additives and preservatives, from non-polluting milk, very small parties and under the order. Labels and bottle shape are reminiscent of home-made milk from grandma.

Design Agency
Nadie Parshina studio (Russia)

Creative Director
Nadie Parshina

Designer
Nadie Parshina

Intriguing Magic

Nothin' Doin'

The Game of Gainful Unemployment is a board game, the object of which is to achieve and sustain happiness without working a nine-to-five job. Players roll dice to move around the board, encountering a variety of situations that increase or decrease their levels of time, money and pleasure. By keeping these three currencies balanced and collecting entrepreneurs, investors, idlers and anarchist mentors, players make their way to the hammock in the center of the board in order to win the game.

Client
Katie Hatz

Design Director
Joe Scorsone

Designer
Katie Hatz

Photography
Brandon Jones

Country
USA

Thymes Offerings

The Thymes Offerings line consists of 8 individually fragranced candles, each crafted to evoke a particular sentiment or emotion. Symbols are created to represent each emotion, and colors are carefully chosen to match each fragrance. Handwriting is used to emphasize the personal nature of gift giving.

Client
Thymes

Design Agency
Zeus Jones (USA)

Design Director
Brad Surcey

Designer
Celeste Prevost

Photography
Tom Matre

Coquette

Coquette is a jewel brand for a luxury jewel and perfume shop in Brussels called "Les précieuses".

Client
Les précieuses

Design Agency
Face to face design (Belgium)

Creative Director
Flore Van Ryn

Designer
Flore Van Ryn

Photography
Benoit Banisse

Jewelry packaging KLOTZ

This packaging is made of six identical wooden cubes made of oiled nutwood. A leather hinge provides the opening function, a paper loop is used as closure. KLOTZ is not only for protecting the jewelry, but also for presenting it! The whole packaging is made out of natural materials and is 100% biodegradable.
The paper sleeve offers excellent opportunities for branding and labeling, and the size of the box creates a larger, more substantial product out of a small piece of jewelry.

Client
self initiated work

Design Agency
kopfloch.at (Austria)

Creative Director
Gerlinde Gruber

Designer
Gerlinde Gruber

Photography
Stephan Friesinger

r¿ng SR118

r¿ng is a 10carat white gold Möbius band with a hand tapped 16gauge internal [female] thread to receive three uncut rough diamonds set individually into screw-in external thread-form propeller claws. It comes housed in a 215-piece hand-painted box with a threaded lid for safekeeping.

Client
In-house Release

Design Agency
Sruli Recht (Iceland)

Creative Director
Sruli Recht

Designer
Sruli Recht

Photography
Marino Thorlacius

Concrete Buckle SR120

Concrete Buckle is a set of 4 hand-cast belt buckles with bar and hook bent from sealed reclaimed steel, boxed with a horse-skin Black Sable, Rose Grey and Bay Red belt.

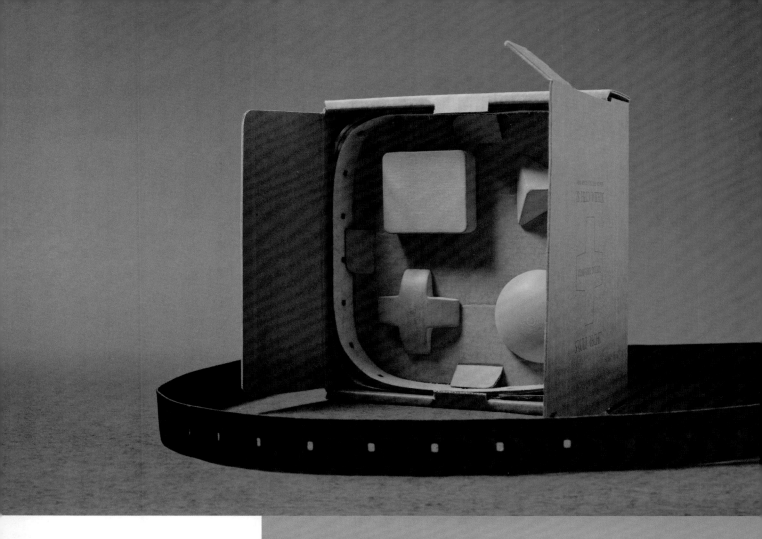

Client
In-house Release

Design Agency
Sruli Recht (Iceland)

Creative Director
Sruli Recht

Designer
Sruli Recht

Photography
Marino Thorlacius

Jordan x Levi's

We worked with Brand Jordan to create the packaging and illustration for the collaboration between Jordan and Levi's 501.

The packaging featured specially designed Jordan and Levi's illustrations and graphics, as well as an illustrated T-shirt and Perspex split box dividers.

Having been sent a pair of pristine Jordans 1s and a fresh pair of Original fit Levi's 501, we went ahead, scalpel in hand, to dissect the shoes and jeans with a tear in our eye! This helped us grip on how each product was made and helped develop an appreciation for the details and construction. We started to design an organic, elegant hand-drawn collage, inspired by individual details from both the products and famous iconic brand elements. We created individual designs for each facet of the box and clear dividers within this. This was a really special project – and one that was great for us to be involved in.

Client
Brand Jordan

Design Agency
ilovedust (UK)

Creative Director
Mark Graham & Ben Beach

Design Director
Ben Beach

Designer
Shan Jiang

Photography
n/a

Reusable packaging

TOT-a-LOT is a brand dedicated to design clothes for twins. After designing the brand, we devised a packaging that conveys the idea of twins and also thought to be reused in the form of drawers for clothes. Having two babies at the same time takes effort and money. TOT-a-LOT is very responsible to the environment so all materials and inks used in their clothes are eco-friendly, so we have designed a simple but impressive packaging focusing on the tone of voice of the brand. Nothing else is needed to do to make it so appealing. The boxes are personalized with the names of babies and parents can begin putting their first clothes each in its corresponding drawer.

Client
Tot-A-Lot

Design Agency
Wesemua (Spain)

Creative Director
Wesemua

Photography
Wesemua

ICA Baby Diapers

The assignment was to create a series of private label baby products that stood out against the A-brands in the shelf for baby products. So, instead of sweet, harmless babies: babies with an attitude! And the funny thing is that babies become even cuter when they are a bit cocky...or obnoxious... or when they are screaming. The photos for ICA's diapers portray babies the way parents are used to see them most of the time...and that is not with a dreamy smile and glittering eyes on a meadow full of flowers! The pattern in the background is created with a look and feel of contemporary wallpaper. It is used as a category design element for all of ICA's baby products. And if you look closely at the characters among the clouds... they got attitude as well!

Client
ICA AB

Design Agency
Designkontoret Silver KB
(Sweden)

Creative Director
Ulf Berlin

Design Director
Cajsa Bratt

Photography
Roland Persson

Lable design for VIS-A-VIS

As the foundation for the hinged label for women's underpants, I wanted to do something bright, feminine and also corresponding with existing packaging logo VIS-A-VIS. Thus born the hinged label in the form of a butterfly. Black and green colors were taken from the box logo VIS-A-VIS to ensure compliance with corporate identity.

Client
VIS-A-VIS

Design Agency:
Vladimir Shmoylov (Russia)

Creative Director
Vladimir Shmoylov

Designer/Photography
Vladimir Shmoylov

FruitFit

This was a concept for a "RePackaging" assignment in which I chose to repackage bras. The packaging was inspired by a KBG commercial in which a young man asked how to tell his girlfriend's breast size. The answer was to simply relate them to a piece of fruit: "Are they apples, oranges, or grapefruits?"
I thought it was a hilariously cute ad and decided to create a line of bras whose packaging corresponded with the appropriate fruit sizes. A cups are apples and B cups are oranges and etc. Since the concept was very "fruity" in nature, the dingy hangers that lingerie normally came with seemed no longer appropriate as I wanted something light and playful.

Client
personal

Design Agency
n/a (USA)

Creative Director
n/a

Design Director
Jennifer Cole Phillips

Designer/Photography
Tiffany Shih

Divine

It was Orrefors, and designer Erika Lagerbielke, who were given the great pleasure of producing the official gift from the people to the Swedish Royal Wedding couple. This exclusive royal set of glasses are not available for the public, but a very similar "retail version" created and named "Divine". How can this set of glasses benefit from the Royal one, without being too ingratiating and finally get a life of its own? With the design and communication concept "Love is Divine", the parallel to the royal set of glasses is rather obvious, as long as the Royal Wedding is on the agenda. When, however, all memories of the wedding have declined, the concept lives on and can easily be transformed into almost anything; Dining is Divine, December is Divine... Always with an appropriate illustration. Neumeister delivered the main concept, packaging design, ads, sales material ; brochure, point of sale material, etc. Huge media attention. Just released.

Client
Orrefors

Design Agency
Neumeister Strategic Design (Sweden)

Creative Director
Peter Neumeister

Design Director
Peter Neumeister

Designer
Carl Larsson

Copywriter
Tor Bergman

Production Manager
Helene Mellander Holm

Farmetone

Farm is a fashion brand and everything they make is really pretty.
The packaging design for the panettones they give away as gifts every Christmas has to be as well.

Their only request was to limit the use of color to one. I created a nice pattern and designed gift tags on the bottom as a special touch.

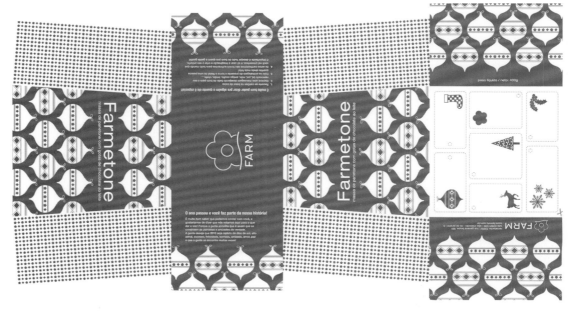

pantone 485

Client
Farm

Designer
Flavia Oliveira

Country
Brazil

Le Pain Quotidien

Various items of packaging and promotional material for bakery chain Le Pain Quotidien in London.

Client
Le Pain Quotidien

Design Agency
Mind Design

Creative Director
Holger Jacobs

Illustrator
Aude Van Ryn

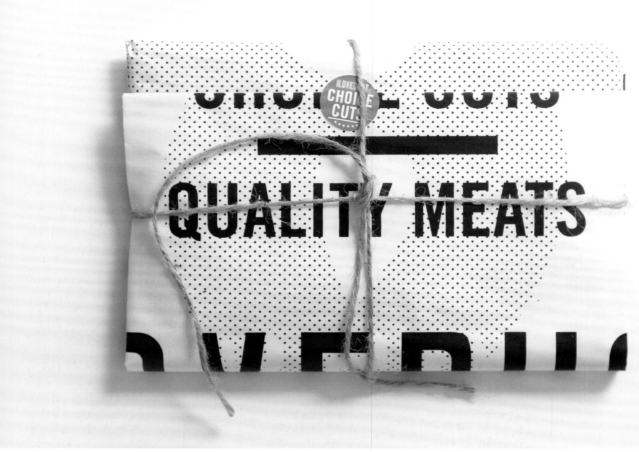

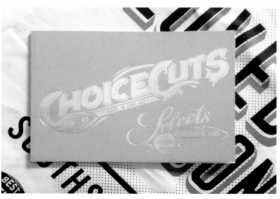
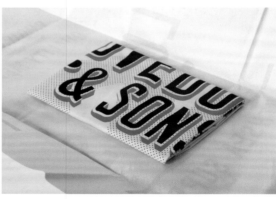

Client
ilovedust

Design Agency
ilovedust (UK)

Creative Director
Mark Graham

Design Director
Johnny Winslade

Designer
Jodie Silsby

Photography
Liz Bick

ilovedust Meat Packing In-house Promo

We recently moved into an old butchers workshop so our shop facing identity naturally began to lend itself to that. We handed screen-printed butchers aprons and tea towels and wrapped those, along with a perfect bound book full of our latest work, in light-weight news print, secured together with twine. These were sent to new and existing clients to showcase our talents – and the response we've had so far has been great!

Client
Self initiated project

Design Agency
Toykyo (Belgium)

Creative Director
Toykyo x Parra

Design Director
Toykyo x Parra

Designer
Parra

Photography
Studio Edelweiss

The Not So Happy Bird

'The Not So Happy Bird' Limited edition porcelain figure designed by Parra & produced by Toykyo in 2010. 2 editions of 25 pieces/ black & white.
size: 22 x 20 x 15 cm

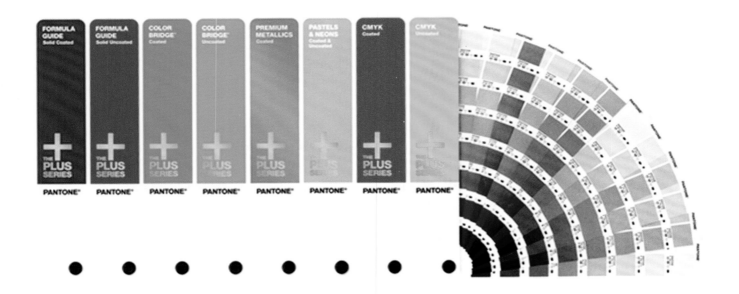

Pantone: The Plus Series

Base's work with Pantone encompasses the identity of the new Pantone Hotel, the 2010 campaign for Pantone's Fashion + Home department, as well as the design of the packaging and campaign for the new Plus Series product line of Pantone's graphics division. For The Plus Series, we've been working with Pantone on the packaging and promotional materials. We started with the name, and worked with Pantone to arrive at "The Plus Series" to emphasize the added benefits of the upgraded product line. Visuals reference the classic Pantone Chip without relying on it. Our design comprises the packaging for the entire Plus Series line—eight fan guides, four chips books, and set boxes—as well as print ads, a series of emails, and a brochure.

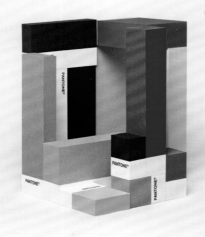

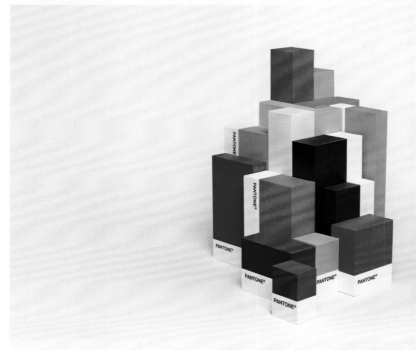

Client
Pantone

Design Agency
Base (USA)

Creative Director
Base

Designer
Base

Photography
Maurice Scheltens
& Liesbeth Abbenes

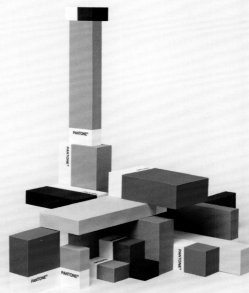

PANTONECANS

My utopian idea was to make sprays with pantone colors inside, the concept was realized with prototypes handmade.
A tribute to the happy combination of design and graffiti art, my passions.

Client
Personal Project

Design Agency
Nico189 (Italy)

Creative Director
Nico Laurora

Designer/Photography
Nico Laurora

Client
Abaco Computers

Design Agency
Nico189 (Italy)

Creative Director
Nico Laurora

Designer/Photography
Nico Laurora

SKATE OR GRAVE

The brief was customize limited edition of Abaco computers for a skate brand.

6-pack Screwdriver Set

This project was done as an exercise to design a more user-friendly package for a set of six screwdrivers. The idea behind the 'open' concept packaging was to allow the consumers to handle the product (screwdrivers) before purchasing them. The package used the construction of a six-pack of beer and applied it to an industrial product. The Mastercraft logo was also developed into a lettermark and wordmark.

Creative Director
Cory Ingwersen

Designer/Photography
Cory Ingwersen

Other
Conceptual project

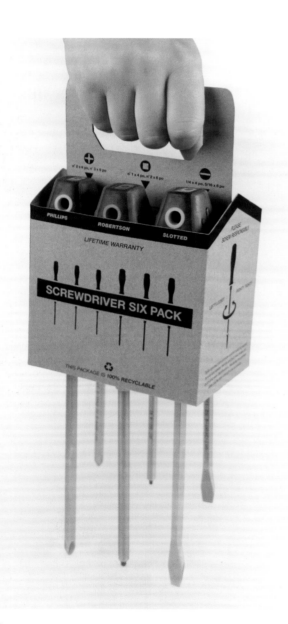
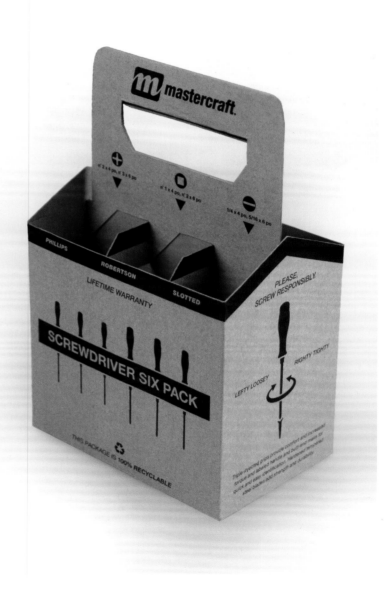

WOW - shoppingbag

WOW—A promotional shopping bag for Bailly Diehl. We were asked by the boutique fashion chain Bailly Diehl in March 2009 to create a paper promotional shopping bag for their planned summer promotion campaign. Their brief was that it should have an unusual motive and should be an "eye catcher." Bearing in mind the 'living accessories' often carried by their target shoppers we decided the use of a pug would project an appropriately sympathetic image. A slit in the mouth area allows the garments purchased to be partially pulled through, giving the impression that the pug is stretching out his tongue, and in this way is participating in the purchase.

Client
bailly diehl

Design Agency
merkwuerdig.com (Germany)

Designer
Nadine häfner, Jennifer staudacher, kai staudacher

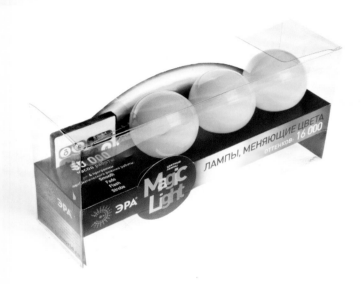
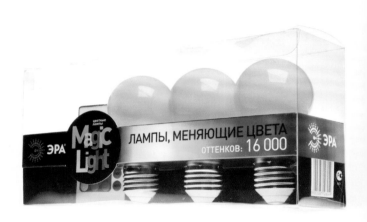

Package design for bulbs change color with remote control

The bulbs change color with remote control. I tried to create packaging that constructively showed product best and graphically hinted that this light bulb changed their color. Packaging consists of a transparent plastic container and cardboard insert with two-sided printing. This insert simply consists in a rigid structure without splicing.

Client
ERA

Design Agency
Vladimir Shmoylov (Russia)

Creative Director
Vladimir Shmoylov

Designer
Vladimir Shmoylov

Photography
ERA

Spatula Putty

We thought of the product as any fast moving consumer good; with our mind towards the creation of a pack that would in fact sell putty.

We created a design that could differentiate the brand (Petrocoll) from its competitors. We employed the female element as our communication platform; a female figure would seem quite out of place at a construction site and would—we hoped—create a buzz among people in the business. Every female figure was dressed up with clothing of various degrees of transparency—the idea being not to be provocative, but to relate the product characteristics (degree of putty overlap) with the respective garment.

Client
Petrocoll Sa

Design Agency
Mouse Graphics (Greece)

Creative Director
Gregory Tsaknakis

Design Director
Aris Pasouris

Illustrator
Ioanna Papaioannou

Graphic Designs for Toothpaste

One of the design objectives was to communicate the idea of an efficient, detailed dental care range, created to solve real problems and, at the same time, differentiate from the leading brands offering a similar image of quality.

We have used a typographic solution, which is very functional because it can be read easily and can communicates clearly the utility of the toothpaste. The overlapping of the letters and their transparencies provide the necessary graphic richness to personalize the range, to show the quality of the products and to suggest the care for details which they have been produced with.

Client
Laboratorios Korott.

Design Agency
Lavernia + Cienfuegos (Spain)

Creative Director
Nacho Lavernia & Alberto Cienfuegos

Designer/Photography
Nacho Lavernia & Alberto Cienfuegos

Vodafone HSDPA Packaging

Client
Vodafone Hungary Zrt.

Design Agency
Café Design (Hungary)

Creative Director
Mr. Attila Simon

Designer
Ms. Gabriella Balázs

Our client is an international telecommunication leader open to innovative solutions.

The task was to design the packaging of a new product: the hsdpa (High-Speed Downlink Packet Access) service. The client's demand was to break with the traditional design style and come up with something brand new to fit the novelty.

Since the former designs were characteristically very illustrative the new design became very clear using few colours only.

Wrapping paper

Wrapping paper for Hunting Lodge, a shop selling designer toys and clothing. The pattern is based on geometric shapes of woolen felt.

Client
Hunting Lodge

Design Agency
Commando Group (Norway)

Creative Director
Commando Group

Designer/Photography
Commando Group

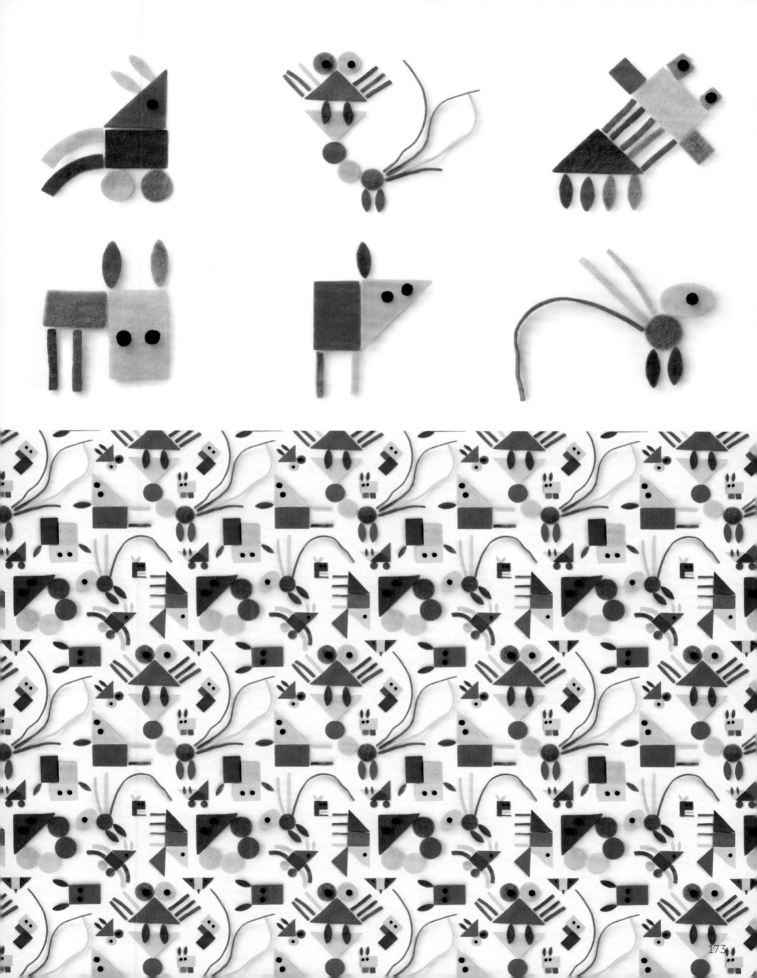

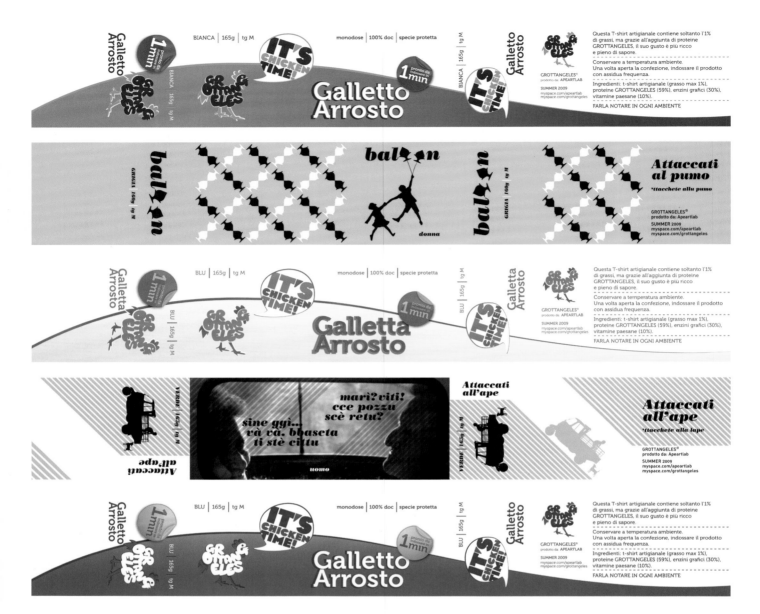

Grottangeles (T-Shirt)

Grottangeles 2009: packaging t-shirt as food items
Grottangeles 2010: packaging t-shirt as ceramic items

Client
Apeartlab (Personal Works)

Design Agency
L-Enfant (Italy)

Creative Director
Pierfrancesco Annicchiarico
And Alessandra Sanarica

Designer
Pierfrancesco Annicchiarico
And Alessandra Sanarica

Photography
Dario Miale

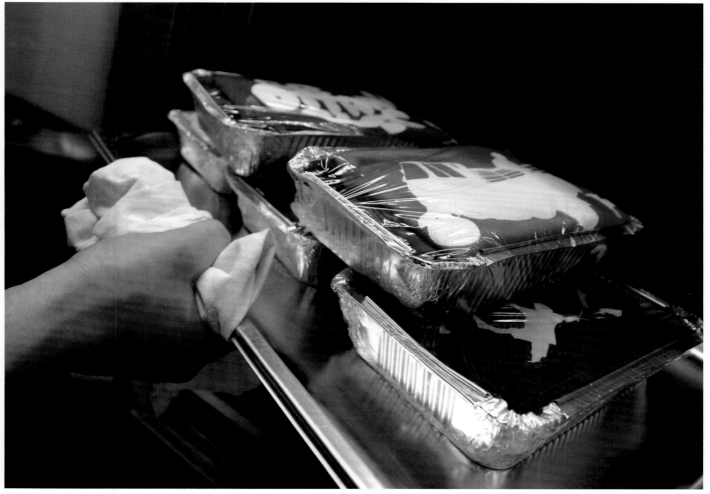

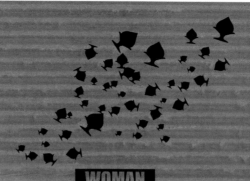

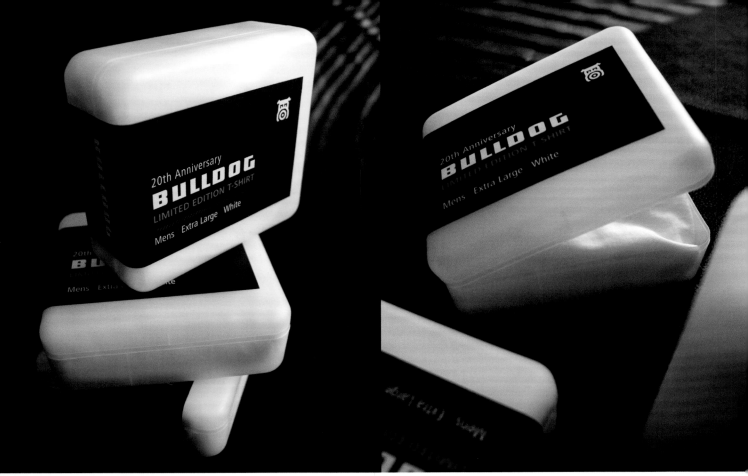

Client
Bulldog

Design Agency
Bulldog (Malta)

Creative Director
Ren Spiteri

Designer
Ren Spiteri

Bulldog 20th Anniversary T-shirts

Over the years, our company has been very comfortable with presenting a witty face to the world – from our choice of name to the imagery that people associate with Bulldog.

We developed this T-shirt to commemorate our 20th anniversary on February 10th, 2010. The piece fuses a crossword puzzle with our logo to create a playful visual short-hand for what Bulldog has done over the last 20 years.

The starting place for this solution was a concise thank-you message from a happy client – the crossword puzzle graphic just seemed to manifest itself. Finding branding related terms to fit in the spaces was rather more challenging.

Bulldog's clients received these limited edition T-shirts, packaged in labelled, PVC containers.

Fleriana Clothes Aromatics

Clothes Aromatics

Client
Provipax Sa

Design Agency
Mouse Graphics (Greece)

Creative Director
Gregory Tsaknakis

Design Director
Gregory Tsaknakis

Illustrator
Ioanna Papaioannou

Music Box

The packaging for an 18 note small music box, with tunes fom Amelie Poulin, Pink Panther or The Rolling Stones.

Graphic Design, Packaging, illustration, Photography & Styling by Two Dot Two.
5.5 cm x 4.5 cm x 2.5 cm

Client
Loja de Estar

Design Agency
Two Dot Two (Portugal)

Designer/Illustrator
Maria Helena Silva

Photography
João Bento Soares

Adaptimals

Adaptimals is a line of stuffed toys portraying animals which have special physical adaptations to facilitate their survival. The gazelle has springs in its legs, the owl's head turns around on a caster, and the hedgehog's elastic drawstring allows it to roll up into a ball to protect it from predators. Each Adaptimal comes in a box covered with fun tidbits of educational information about its species.

Client
Katie Hatz

Design Director
Kelly Holohan

Designer
Katie Hatz

Photography
Brandon Jones

Country
USA

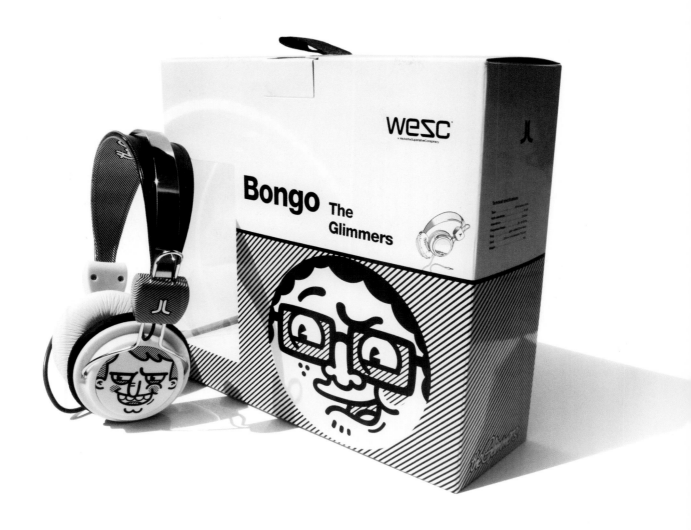

Glimmers Bongo

The Glimmers' Bongo headphone.
Limited edition headphone for the Belgian dj duo 'The Glimmers'.
Designed by Pointdextr for Toykyo studio in 2009.

Client
Wesc

Design Agency
Toykyo (Belgium)

Creative Director
Pointdextr

Designer
Pointdextr

Photography
Benjamin Van Oost for Toykyo

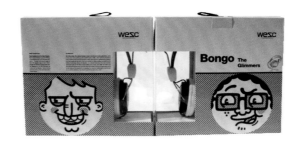

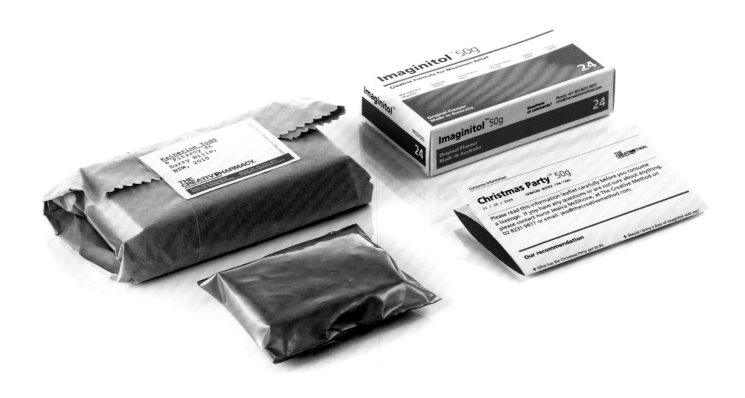

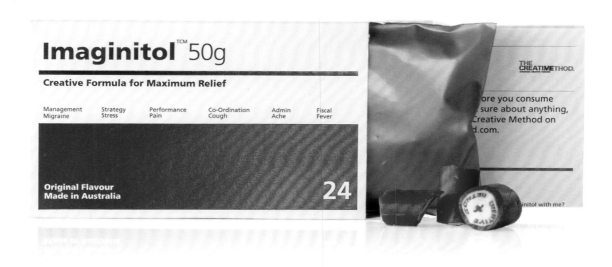

Imaginitol

The brief was to create an interesting and engaging invitation to The Creative Method Xmas party. It needed to illustrate what we did and to create a high level of interest and anticipation for the party. It needed to be humorous and memorable. It was also required to work as a new business piece outside of the Christmas invitation.

We based the idea on an imaginary pharmaceutical tablet that would solve their creative issues. Initially they were emailed as doctors' prescription, followed by the package in a discrete paper bag. The invitation and the tablets were located inside. The party included staff dressed as doctors & medicinal shots administered by transvestites. The box and invitation were used as a new business teaser.

Client
The Creative Method

Design Agency
The Creative Method (Australia)

Creative Director
Tony Ibbotson

Designer
Mayra Monobe & Sinead McDevitt

Digestive Wellness

Brand identity and package design for a complete line of digestive care products.

Client
East Meets West Nutrition

Design Agency
Christian Hanson (Canada)

Creative Director
Christian Hanson

Designer
Christian Hanson

Women' Secret sun care products

Just in time for summer, Women'Secret has launched a line of sun care products. An extension of the labels w'eau, the line features three products: sun lotion, after-sun cream, and facial sun cream. We've designed the packages for the set, using clear, minimal graphics and color so that nothing gets in the way of your rays.

Client
Women' Secret

Design Agency
Base (Spain)

Creative Director
Base

Designer/Photography
Base

Beautifully Delicious

Delicious is designing a logo that beautifully captures the essence of a brand so simply and effectively it can help enable the brand grow across a diverse range of over 25 tasty beauty products.

Client
KMI Brands

Design Agency
R Design (UK)

Creative Director
Dave Richmond

Designer
Charlotte Hayes

Avril Lavigne Fragrance Packaging Design

The concept of the fragrance bottle and the secondary packaging for "Black Star" (first fragrance of Avril Lavigne) reflects her personality and image of this particular point in her career. She states that the fragrance is herself in a bottle. The similarities between the packaging and her appearance are pink, black, stars and studs, femininity with punk.

Client
P&G (Procter & Gamble)

Design Agency
Bond. And SelectNY in New York (USA)

Creative Director
Matthias Kaeding

Designer/Photography
Matthias Kaeding

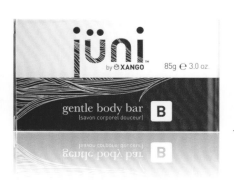

JUNI

Client
XanGo

Design Agency
ONLY Creatives (USA)

Creative Director
Chris Metcalfe

Photography
Ash Ram

Juni is an all-organic product line for hair and body formulated for gentle yet deep cleansing for the entire family. We started by employing vibrant colors that were indigenous to nature yet also conveyed a feeling of fresh organic cleanliness. The colors also served to identify different sections of the product family with the green family representing 'hair' and the citrus orange and reds representing 'body'. The usage of the hair pattern was to establish a subtle yet direct association with the hair and body as well as to create a unique organic texture. Overall, we sought to create a design that was simple and natural as the product it contains.

Oriflame Very Me

What has shoes and lipsticks to do with each other? Well they are both girls' best friends.

When creating a label for cosmetics targeting young female adults, you have to dive into their world of accessories. Since both shoes and lipsticks are important things in the daily life of young women, why don't bring them together? Working with print on a transparent packing gives the package a depth, elegance and modernity. but still to an affordable price, which is necessary when targeting young people. The packaging also highlights the colour of the product and makes it easy to out pick your favourite.

Client
Oriflame Cosmetics

Design Agency
Designkontoret Silver KB (Sweden)

Creative Director
André Hindersson

Design Director
Cajsa Bratt

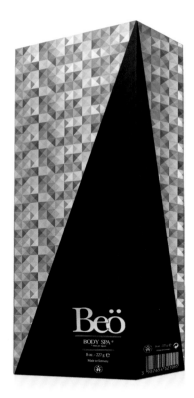
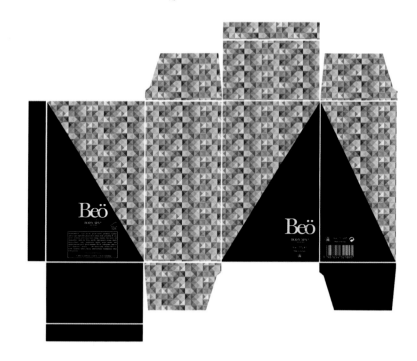

Client
2B4W (new company developing)

Design Agency
Pikartzo (France)

Creative Director
Alexandre MANET

Design Director
Alexandre MANET

Designer/Photography
Pikartzo

Beö

Primarily designed for sunny months, BEÖ was designed as a dual-function product. The colorful geometric patterns symbolizing the madness of the summer, they represent the mood of joy and delusions of all kinds in the existing makeshift barbecue, diving into the pool imposed by heat and laughter shared with friends. Often during these summer days, holidays and rest are synonymous. Matte black as for him, represents the classy side and high-end product. Between an afternoon and evening sun in a night club, there are many differences. However Beo may be the product adequately to any of these situations.

Client
RNB Laboratorios

Design Agency
Lavernia + Cienfuegos (Spain)

Creative Director
Nacho Lavernia
& Alberto Cienfuegos

Designer/Photography
Nacho Lavernia
& Alberto Cienfuegos

Codizia Man: Packaging for Men's Fragrance

Codizia Man is a men's fragrance from the same brand launched three years ago for the female market. It shares the same product quality, positioning and differences, and has a much lower price than high-end colognes.

The packaging communicates similar attributes of sensuality, elegance and dynamism. It follows the same language and some characteristics of its female predecessor, as in the solution for the join between body and cap. But it changes to reaffirm the male personality for example in colour and volume, moving from the horizontal to the vertical.

It is distributed exclusively through the Mercadona supermarket chain.ts.

Rizon Spa

The spa is right by the sea, and the name "rizon" is from the word "horizon." They use seaweeds for their spa products. A combination of black and white photographs and illustrations of seaweeds is used for the packaging and the invitation.

Client
Rizon Spa

Design Agency
Eulie Lee Design (USA)

Designer
Eulie Lee

SLA Paris | Restyling of brand and packs

Restyling of brand and packaging for cosmetic products.

Client
Mpbata

Design Agency
Zoo Studio (Spain)

Creative Director
Jordi Vila

Designer
Xavier Castells

Photography
Oscar Araujo

Scent Stories

Perfume packaging design and the concept of the perfume were always our dream project. So we took men's fragrance as our challenge.

At the beginning we were concentrating on the idea of the scent itself. We found inspiration in the great, dark literature and distinctive, strong characters. We tried to describe the dark sides of men's nature with line of scents named after famous writers.

We packed the scents into bottles which resemble both old glass perfume bottles and the classic shape of the inkwell. We made them white, added black strong lettering and heads of characters which loosely recalled the author's famous masterpieces.

Client
Ah&Oh studio

Design Agency
Ah&Oh studio (Poland)

Creative Director
Magdalena Kalek, Kamil Jerzykowski

Designer
Magdalena Kalek, Kamil Jerzykowski

Green and Spring

The design needs to capture the magic and uniqueness of the product and focuses on the countryside and namely the birds that can be spotted within the British Isles.

Each product, within the three ranges, is characterised by an individual British bird and segmented by individual colour palettes that are directly linked to the emotions associated with them. The 'Indulging' products are inspired by warming hues of reds, pinks and oranges, the 'Relaxing' products focus on softer and more romantic shades such as purples and greys with the 'Revitalising' range engaging the senses with zesty yellows and greens. The front of pack copy forms part of the design story and highlights the key ingredients and their benefits for each product, bringing a 'poetic clarity' to the overall visual world of Green and Spring.

Client
Green and Spring

Design Agency
Pearlfisher (UK)

Creative Director
Natalie Chung

Design Director
Sarah Butler

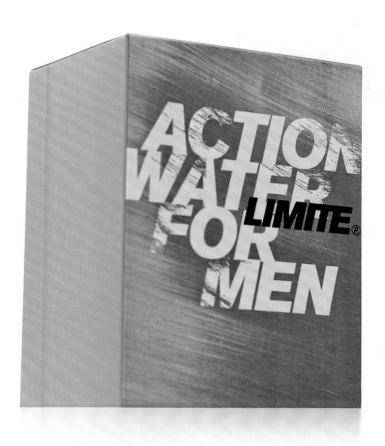

Limite: Packaging for Men's Fragrance

LIMITE cologne is designed for a young, dynamic, active, urban public. Targeting at the sports cologne category, it required a standard pack at the lowest possible cost. We decided to make a virtue of necessity and eliminate the cap, which was a significant cost component. We replaced it with a safety clip that prevented the spraying and was a simple injection molded flat piece of plastic, perforated with the brand name to convert it into the key customizing element. The graphics, color, typography and texture used for the text on the box reinforced its masculine, urban and sporty character.

Client
RNB Laboratorios

Design Agency
Lavernia + Cienfuegos (Spain)

Creative Director
Nacho Lavernia & Alberto Cienfuegos

Designer/Photography
Nacho Lavernia & Alberto Cienfuegos

Little Me

A re-brand for a popular range of organic baby toiletries made from organic ingredients that contain gentle formulations and avoid synthetic fragrances. This will be found in over 300 stores across the UK. The brief was to re-design the packs to appeal to mothers and to stand out in an extremely competitive market place, whilst communicating the organic benefits and origins of the product. We achieved this by using a large decorative floral illustration incorporating the brand identity with the use of bold, bright colours, which emulate children's book illustrations.

Client
KMI Brands

Design Agency
R Design (UK)

Creative Director
Dave Richmond

Designer
Charlotte Hayes

Pure and Nature

The Pure and Nature product group is a Kruidvat* (A.S. Watson - Health & Beauty Continental Europe) private-label range designed for daily face and body skincare. Pure and Nature products consist of all-natural ingredients, and are the first private-label products in the Netherlands to carry the BIDH quality mark. (The BIDH mark guarantees that products do not contain any synthetic coloring, scent or preservatives, as well as any synthetic silicones, paraffin or other oil-based products. Furthermore, the natural ingredients have been, whenever possible organically cultivated and have not been tested on animals).

Reggs created a design that is eye-catching on the shelves, but which also radiates a simple, uncomplicated look-and-feel that reflects and emphasizes the products' all-natural, high-quality characteristics. The Pure and Nature packaging makes an appealing, fresh and pure statement, in which the Kruidvat values also remain visible.

Client
A.S. Watson

Design Agency
Reggs (The Netherlands)

Creative Director
Léonie van Dorssen

Design Director
Lian van Meerendonk

Designer
Léonie van Dorssen

Photography
Ingemar Paalman

Asira Amenities Packaging Design

Radissons in-hotel amenity brand "Asira" needed an update in 2010 after the visual identity of the Hotel Group was changed in 2009. The goal was to create a contemporary design with perfect readability/usability for tired travelers. The image of "Asira" is fresh and vibrant.

Client
Radisson Hotels and Resorts,
Carlson Hotel Group

Design Agency
Bond. And SelectNY in New York
(USA)

Creative Director
Matthias Kaeding

Designer/Photography
Matthias Kaeding

Shu Uemura brand film

We Are Plus collaborated with the team at Shu Uemura to create an inspirational brand film that celebrates their new product lines. We created an upbeat and eye-catching film to explore and celebrate their annual collections.

Client
Shu Uemura

Design Agency
We Are Plus (USA)

Creative Director
Jeremy Hollister / Judy Wellfare

Designers
Judy Wellfare, Laura Reiland, Jeremy Hollister

Photography
Shu Uemura

Aco Spotless

ACO Spotless, products that are made for Spotless days. The assignment for this project was to create a skin care series for young skin. With the right tone of voice to communicate in a relaxed way what the users wanted to achieve "Spotless".

The combination of strong colours and technical packaging gives a whole that is different in the category. Apart from the straight on design, even the product promise made with an attitude that signaled what the product helped the consumer with.

Assignment included design strategy and copy that supported a complete system of products that both treated and prevented spots. ACO Spotless are only sold in the pharmacies in the Nordic countries.

Client
Aco Skin Nordic AB

Design Agency
Designkontoret Silver KB
(Sweden)

Creative Director
André Hindersson

Designer Blanca Waller,
Monica Holm

Aquatus+ Abstract and Nature Edition

This package design was made for the hungarian hygienic products, that will help women to improve their quality of life and can be used for prevention at the same time. It is a really interesting and sensitive task. The package is stylish and feminine.

Client:
Intimpharma Ltd.

Design Agency:
Fontos Graphic Design Studio (Hungary)

Creative Director:
Mate Olah

Designer:
Mate Olah

Other:
3D modelling by Gabor Gloviczki

Belmacz Beauty

Packaging and promotional material for the Belmacz Beauty range. Products include a lip gloss with real gold leaf and a powder with crushed real pearl. The packaging makes use of different textures, print and embossing effects.

Client
Belmacz

Design Agency
Mind Design

Creative Director
Holger Jacobs

Photography
Franck Allais (for product images)

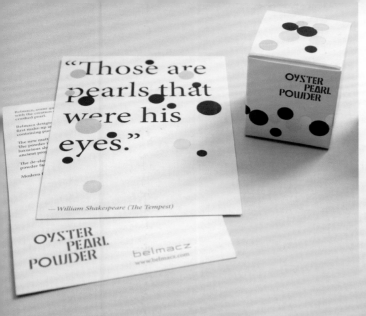
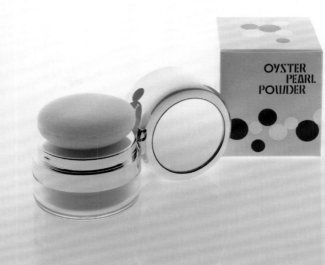

Shengaia Origami Packaging

To compliment their range of apothecary and massage treatments, we created a sumptuous brand identity that evoked nature and a blend of East and West through a rich palette of refined colours and material finishes. We also developed bespoke packaging solutions for a range of ten skincare products for Shengaia, each with a different opening mechanism to epitomize their luxurious brand ethos.

Client
Shengaia Skincare

Design Agency
Campbell Hay (England)

Creative Director
Charlie Hay

Designer
Charlie Hay

Photography
Yann Binet

Aveeno Positively Ageless™ Lifting and Firming Facial Care

With a sustainability initiative firmly in place, Johnson & Johnson GSDO partnered with Hoffman Creative on the packaging design for their premium facial care line, Positively Ageless™ Lifting and Firming. The three SKU family needed a responsible solution to its secondary package that both reflected J&J's ecological commitment (eg. reduced packaging footprint) and Aveeno's brand position on natural skin care. Sourcing 3-sided, e-flute paper liners and a plastic sleeve with post-consumer recycled content, Hoffman Creative was able to design a package that allowed light to flow onto the primary structure while delivering a compelling environmental story instantly to consumers on-shelf.

Client
Johnson & Johnson GSDO

Design Agency
Hoffman Creative (USA)

Creative Director
Eric Hoffman

Photography
Timothy Hogan

Aveeno Positively Nourishing™ Body Care

For the launch of a new, scented body care offering by Aveeno, Hoffman Creative partnered with Johnson & Johnson GSDO to develop graphic treatments and a fresh color palette for all four SKU's in the Positively Nourishing™ family. Balancing between the brand's heritage and the desire to appeal to a younger audience, Hoffman Creative enlisted the help of artist, Abigail Borg, to sketch botanical illustrations that infused a modern hand with the traditional sensibility of still-life rendering.

Client
Johnson & Johnson GSDO

Design Agency
Hoffman Creative (USA)

Creative Director
Eric Hoffman

Art Director
Charlotte Pao

Photography
Timothy Hogan

Aveeno® Therapeutic Bath Formulas

Johnson & Johnson GSDO partnered with Hoffman Creative for a graphics and materials refresh for Aveeno's historic bath offerings including powdered and liquid oatmeal formulas. By building an aspirational color palette, updating the graphics hierarchy, introducing contemporary illustrations and utilizing more experiential materials, Aveeno's hallmark product line had significant differentiation from competing store tribute brands.

Client
Johnson & Johnson GSDO

Design Agency
Hoffman Creative (USA)

Creative Director
Eric Hoffman

Photography
Timothy Hogan

M Picaut

M Picaut is a new exclusive Swedish skin care company with products that combine anti-age with natural ingredients. We have designed the identity, packaging design and web site. Glossy surfaces and generous white areas are communicating luxury and high-tech while the orange colour announces the Backthorn series.

Client
Mette Picaut AB

Design Agency
OCH (Sweden)

Creative Director
Johanna Kallin & Magnus Engström

Designer
Johanna Kallin & Magnus Engström

Photography
Kalle Sanner

FOUND SKINCARE PACKAGING

Enigmatic venturer Éric Cosson came to Campbell Hay with an idea for a range of high-end skincare products. They would be made using effective concentrations of precious and unusual ingredients, such as oyster extract and snake venom, sourced from all over the world.

Together we developed a brand that reflected the high-tech exoticism of Éric's concept, and a range of packaging to clearly express the unique properties of each product.

Client
Found

Design Agency
Campbell Hay (England)

Creative Director
Charlie Hay

Designer
Charlie Hay

Photography
Yann Binet

Dizzying Magic

Gyulai Pálinka

Client
Gyulai Pálinka Manufactory

Design Agency
Café Design (Hungary)

Creative Director
Mr. Attila Simon

Design Director
Mr. Péter Berki

The original 'pálinka' is a distilled spirit made of fermented fruits. Only those products, which are made from Hungarian fruits, can be referred to as 'pálinka'. They have 100% fruit content and have got 37,5% of alcohol strength as a minimum.

The first distillery in Gyula was founded by János György Harruckern in 1731. This nearly 300 year-tradition is carried on by Gyulai Pálinka Manufactory these days.

The task is to create a new packaging design which focuses on the mature young generation. The key motive of the new design is the Hungarian hound, symbolizing fidelity, nobility and all traditional values.

IPACS Winery

Ipacs Winery is a young, emerging, dynamic developing and already successful winery led by an ambitious owner.
Our task was to design the CI and the packaging for the winery. Neither the package-technology nor the budget of Ipacs could rival with the historic wineries, thus the designer's goal was to take advantage of the disadvantage. The quite unusual flag-like label was born thanks to this pressure. The label holds all important information so the tag on the bottle can concentrate exclusively on the branding.

Client
Ipacs Winery Ltd. – Ipacs Pince Kft.

Design Agency
Café Design (Hungary)

Creative Director
Mr. Attila Simon

Design Director
Mr. Tamás Veress

Original Vodka & Gin

Product and label design for the Swedish Original Vodka and Gin brand.

Client
AB Åbro Bryggeri / AB Åbro Breweries Sweden

Design Agency
Morris Pinewood Stockholm (Sweden)

Creative Director
Mattias Frodlund

Designer
Mattias Frodlund

Client
Bauer Vigneron Blienchschwiller - Alsace

Design Agency
Pikartzo (France)

Creative Director
Alexandre MANET

Designer/Photography
Pikartzo

Win(e)

Born from the desire to create an original product, Win (e) is intended primarily for Businessmen and chic and trendy places. With a tie-shaped label on the front tie and a bow tie on the back, the graphics aim directly at the customer. A trademark is a word game, giving a sense of victory, the color code chosen, contains the features related to purple. Indeed purple is the color for dreamers, who want to calm some emotions, to restrain anger or distress. Win (e) makes it possible to escape from everyday life, and relieve the stress of quotidian.

The box uses the same graphic; the tie is transparent, allowing the product to appear through.

Finca Nueva

Wine bottles designed for Finca Nueva, which means pleasure wine.

Client
BODEGAS FINCA NUEVA

Design Agency
(calcco) Comunicación Visual (Spain)

Creative Director
Raúl Barrio, Sergio Aja

Design Director
Raúl Barrio

Designer
Sergio Larrauri

Photography
Sergio Aja

Hechura

Wine bottles designed for HECHURA

Client
BODEGAS RAMON BILBAO

Design Agency
(calcco) Comunicación Visual (Spain)

Creative Director
Sergio Aja

Design Director
Raúl Barrio

Designer
Raúl Barrio

Photography
Sergio Aja

MAYRAH WINE

The wine is from Australia, and "mayrah" means "spring" in the aboriginal language. The animals are jumping up because they are so happy that spring has finally come.

Client
MAYRAH WINE (USA)

Design Agency
Eulie Lee Design

Designer
Eulie Lee

Client
Berntson Vin AB

Design Agency
Neumeister Strategic Design (Sweden)

Creative Director
Peter Neumeister

Design Director
Peter Neumeister

Designer
Tobias Andersson

Copywriter
Tor Bergman

Paul Sapin

When Paul Sapin launched their series of wine, white, red and rosé, in PET-bottles the awaited success did not occur – although reviews were good. What was the problem? Neumeister was approached to take a look at the packaging design. "Easy to go", is the main idea. Classic design combines with a modern approach. Old meets new, in the same way as the product; traditional French wine, produced in a traditional way, but with a whole new packaging concept. And the twisted label as an eye catcher.

Spanish White Guerrilla

Maetierra's Spanish White Guerrilla is a set of eight single-varietal wines made from the world's most prestigious white grapes (Albariño, Gewürztraminer, Riesling, Sauvignon Blanc, Verdejo, Chardonnay and Viognier), being grown in La Rioja for the first time. This set, as its design, is a revolution in itself. The choice of picture (by Brosmind Studio) as a technique to develop the creative work was based on the idea that the set itself was the embodiment of that revolution, through those guerrilla-wines, each representing their variety and development in a land of internationally renowned wine-making tradition.

Finally, the richness in details of each character, with a fair dose of fantasy and humor, has provided to each wine the necessary elements to communicate and intensify the revolutionary character that inspires Spanish White Guerrilla.

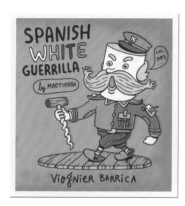

Client
Vintae

Design Agency
Moruba (Spain)

Creative Director
Daniel Morales / Javier Euba

Designer
Daniel Morales / Javier Euba

Illustrator
Brosmind Studio

Build Your Own

The aim was to create a unique gift to give our clients at Christmas and to act as a new business introduction. It needed to remind them of whom we were and the many hours that we had put into our work. It needed to feature all of our staff, to reflect our creativity and sense of humor. We obtained high quality cleanskin wines and created our own labels. Each label was based on one staff member. It included a number of facial features and the client was encouraged to BYO—Build Their Own. The wine and the label are the perfect substitutes when the real thing cannot be there.

Client
The Creative Method

Design Agency
The Creative Method (Australia)

Creative Director
Tony Ibbotson

Designer
Andi Yanto

Beach House Wines

Concept rebranding for Beach House, a fine wine company. The rebrand included designing a new logo and creating new bottles. The clean design was complimented by the striking white bottles.

Client
Beach House

Design Agency
Steph Baxter (New Zealand)

Designer
Steph Baxter

Photography
Tony Brownjohn

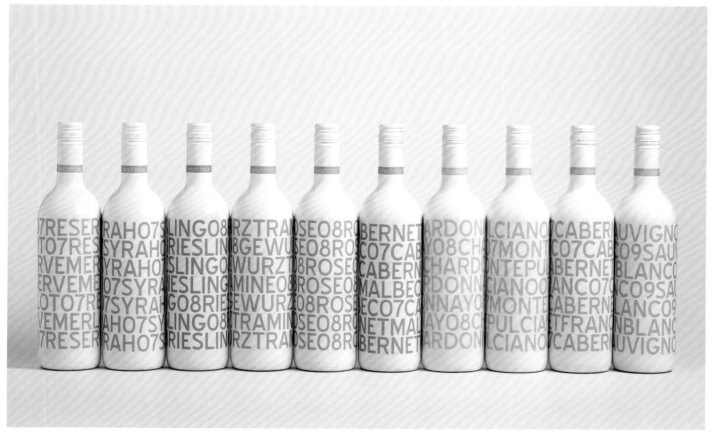

DAME VIDA

Wine bottle designed for DAME VIDA.

Client
DEL VINO AL AGUA S.L.

Design Agency
(calcco) Comunicación Visual (Spain)

Creative Director
Raúl Barrio

Designer
Raúl Barrio

Photography
Sergio Aja

Client
Bodegas Lan

Design Agency
(calcco) Comunicación Visual (Spain)

Creative Director
Sergio Aja

Design Director
Raúl Barrio

Designer
Eduardo Del Pozo

Photography
Sergio Aja

Concertum

Wine bottle designed for Concertum.

Estate Range

Client
Longview Vineyard

Design Agency
Voice (Australia)

Creative Director
Anthony De Leo, Scott Carslake

Design Director
Anthony De Leo, Scott Carslake

Designer
Anthony De Leo

Longview's Estate range has been available for 10 years. After the vineyard was sold, the new owners commissioned the redesign of the original labels with the specific purpose of increasing sales and portraying a message of quality and uniqueness. Targeting 25 to 45 year olds, the range is priced from $20 to $45 AUD and available in some international markets, primarily Australia.

Taking the diverse names of the wines, illustrations were created to resemble each name with a recognisable, related graphic image. Typography, stock and illustration style were all characteristic of the hands on, contemporary nature of the vineyard; an aesthetic used to communicate the personal attention of each product to the market.

Snälleröds

Snälleröds was established in 1901, but fell in to oblivion until Eric Berntson re-launched the brand with a series of organic liquor products. Our mission was to establish the history of Snälleröds with a distinct brand platform and also to invent equally eco-friendly design solutions. The history of Snälleröds is apparent in every aspect of its design and communication with the tagline "A true story" and visual solutions always telling that very story. The packaging design, in some cases, is the first ever of its kind; everything being totally environmentally friendly. Snälleröds has received a lot of extremely positive media attention. The innovative packaging design and the appealing gift packages has been as important for the buzz as the actual products.

Client
Bertson Brands

Design Agency
Neumeister Strategic Design (Sweden)

Creative Director
Peter Neumeister

Design Director
Peter Neumeister

Designer
Tobias Andersson

Production Manager
Helene Mellander Holm

Good Wine Company

We were approached by the Good Wine Company who wanted a completely new concept for their wine bottle design. They asked us to rebrand their new range of organic, and wanted a fresh, clean graphic to reflect the company's organic and natural ethos. We also designed and produced exclusive box packing, made from 100% recycled stock to reinforce the sustainable nature of the product and the company.

Client
Good Wine Company

Design Agency
ilovedust (UK)

Creative Director
Mark Graham

Design Director
Johnny Winslade

Designer
Johnny Winslade

Photography
n/a

Wine packaging

The client started winery only a couple of years ago. This is not his main profession, rather a love of his life. The designer's task was to create a logo for the first vintage to go on the market. The challenge of this task was to create not only a logo for a product but to give a design to define the whole motivation behind the product.

An elegant logo was expected that stood out from the average line of wine logos. The final creation is a cluster of grapes made of numbers.

Client
Györök Pince – Györök Winery

Design Agency
Café Design (Hungary)

Creative Director
Mr. Attila Simon

Design Director
Mr. Tamás Veress

Backyard Vinyards

I have been contacted by Laurie Millotte to do an illustration for Backyard vineyards... The winery is located just outside of metro Vancouver, BC, Canada, hence the name and the illustration! Brandever always has amazing concepts behind their work, and I am so pleased to have participated in one of them...

Client
Backyard Vinyards

Design Agency
Brandever (Canada)

Creative Director
Laurie Millotte & Bernie Hadley-Beauregard

Design Director
Laurie Millotte

Designer
Fabien Barral

Photography
Fabien Barral & Laurie Millotte

Merula, wine packaging

Client
Merula

Design Agency
Base (Spain)

Creative Director
Base

Designer/Photography
Base

In the village of Torrelles de Foix is a plot of grape-growing land known as Terrer de Sapera. From this land comes the new wine on the block, a merlot called Merula. The name Merula means blackbird in Latin—mirlo, also the source of the name of the merlot grape, whose dark wine is said to resemble the color of blackbird feathers. We designed the label, capsule, and a six-bottle case for Merula. As it's a new winery producing a grape not typical in Alt Penedes, we opted for a non-traditional approach. Building on the name Merula, we worked with illustrator Miguel Ordoñez to create a friendly blackbird.

Fygein Adynaton

The title "Fygein adynaton" comes from the ancient Greek maxim "Pepromenon phygein adynaton", meaning that fate cannot be avoided.

A drop of wine, that most people would just wipe off the neck of the bottle, inspired us to create this design —which proves our predilection for detail.

Having served the wine, you will notice that a drop flows down the neck of the bottle, towards the point where three parallel lines are drawn, one under the other. The line, near which the drop will stop, is the one that foretells your future.

To discover what the future holds for you, read the back of the bottle the prediction corresponding to that particular line and drink to its coming true!

Client
Chris Trivizas (Self Promotion)

Design Agency
Chris Trivizas Design (Greece)

Creative Director
Chris Trivizas

Designer
Chris Trivizas

2010

2010 is the name of a wine to share with friends and celebrate the beginning of the new year. This product has been elaborated using organic farming methods without herbicides, pesticides or aggressive chemical filtrations. It's a limited numbered edition of 2010 bottles. The label wraps the bottle with the help of a rubber band without using adhesives. Once the wine is finished, the bottle can be returned and the label has also a second life as a poster with this message about 2010: "Got a feeling it's gonna be a good year". Cheers!

Client
Casa Mariol

Design Agency
Bendita Gloria (Spain)

Creative Director
Bendita Gloria

Designer
Bendita Gloria

Karadag wine

Design Agency
Nadie Parshina studio (Russia)

Creative Director
Nadie Parshina

Designer
Nadie Parshina

Karadag wine now is just a concept. The main idea of a collection is to remind on the disappearing breeds of animals and birds which live in Karadag. Karadag is a conservancy area on the Pontic Sea, in the Crimea. Each bottle describes an animal or a bird. The part of money from wine sale is listed in WWF.

Client
Sipsmith

Design Agency
Big Fish ® (UK)

Creative Director
Perry Haydn Taylor

Design Director
Victoria Sawdon

Designer
Victoria Sawdon

Photography
Big Fish ®

SIPSMITH

Create a super-premium brand for the first new London distillery to be granted a licence in the last 190 years. The brand identity and packaging were created as a homage to their finely crafted small-batch process, the beauty of their swan neck copper still, and their approach, which was to respect the old and embrace the new.

MATSU

Client
Vintae

Design Agency
Moruba (Spain)

Creative Director
Daniel Morales / Javier Euba

Designer
Daniel Morales / Javier Euba

Photography
Bèla Adler & Salvador Fresneda

The solution adopted is faithful to Matsu's philosophy, his image has been stripped from all sorts of tricks to link directly with nature and with the people who cares about it. Thus, the Matsu's wine triology, "El Pícaro", "El Recio" and "El Viejo" are represented by a portrait series of three generations that devote their lives to the field. Each one's personality embodies the characteristics of the wine that gets its name.

The Kraken

The spiced Caribbean rum market is dominated by a few old mainstream brands and there was a real opportunity to create a product with a little more attitude to target the younger segmentation of this growing market. The king of the seas is the mythical Kraken which has been brought into the vernacular by Johnny Depp's pirate movie franchise. It lends itself well to the old engraving styles and the same era inspired the bespoke loop necked bottle.

Client
Proximo Spirits

Design Agency
Stranger and Stranger Ltd
(UK)

Creative Director
Kevin Shaw

Designer
Kevin Shaw

Spice Tree

The Spice Tree pack features a Klimt inspired golden tree to reflect the complex flavors derived from the unique wooden barrels, which give this whisky its name. Spice Tree is an unusual malt whisky made with extra toasted barrel staves for a flavor described as psychedelic elegance. The pack needed to reflect traditional whisky craft with a touch of wild originality.

Client
Compass Box Whisky Co.

Design Agency
Stranger and Stranger Ltd
(United Kingdom)

Creative Director
Kevin Shaw

Design Directo
Kevin Shaw, Guy Pratt

Designer
Guy Pratt

Casa Mariol wine collection

Casa Mariol is a family-owned winery that has been elaborating wines in Terra Alta for over one hundred years. They have confidence in their agricultural model and consider their environmental commitment as something natural and deeply rooted in their everyday activity. Casa Mariol defends what is natural in the broadest sense of the word and is not at ease with the luxury that often goes together with the wine industry. For example, the bottles clearly call a spade a spade, by their grape variety and their age avoiding romantic cheesy names. Mariol makes homemade wines and even the design has also been resolved using homemade tools such as Wordart, Excel and Cliparts.

Client
Casa Mariol

Design Agency
Bendita Gloria (Spain)

Creative Director
Bendita Gloria

Designer
Bendita Gloria

Sunboom wine

Sunboom wine is a concept. The main idea is that grapes for this wine grow in a sunny district. Wine has absorbed so much solar beams that it tastes like sun. Manufacturers of this wine have invited famous writers to try the wine and describe their sensations and thoughts. All their notes have transferred on bottles without changes.

Design Agency
Nadie Parshina studio (Russia)

Creative Director
Nadie Parshina

Designer
Nadie Parshina

Client
Pernod Ricard Nordic

Design Agency
Designkontoret Silver KB (Sweden)

Creative Director
André Hindersson

Designer
André Hindersson

V&S Wine Maru

Maru is Japanese for circle. The balance that is inherent in the form of the circle is reflected in this Alsace wine, which V&S Group has developed as an accompaniment to sushi. The product name is a part of the design commission, which has been nominated in the European EPICA awards.

Flirtini wine

This is a cranberry wine targeting at girls. The bottle and the label reinforce the feminine and eco-friendly components of this wine.

Design Agency
DDH studio (Russia)

Creative Director
Nadie Parshina

Designer
Nadie Parshina

Liqueurs

In this product, following the same idea of the brand Feitoàmão® by Boa Boca Gourmet® we use the bold colors and clean typography. We choose a bottle (I didn't design that), and I try to conceive a elegant box. This box won a Sena da Silva Award 2009 from the Portuguese Design Center.

Client
Boa Boca Gourmet

Design Agency
António João Policarpo
/ Policarpo Design (Portugal)

Design Director
António João Policarpo

Designer/Photography
António João Policarpo

Back Label

Wineries will sell cleanskins to dump excess or unwanted wine stocks and to avoid the negative consequences of discounting their existing brands by doing so. With a price point of $5.99 AUD, Back Label wine competes in the cleanskin wine market. Due to the price point we were required to work with an extremely limited design and production budget.

The solution works on 2 levels, both of which utilize the brand name. As a front label, Back Label appears backwards, however when the bottle is rotated and the label becomes a back label, the brand name appears correct. As the wine is consumed, Back Label becomes clearer as the magnification decreases.

Client
Back Label

Design Agency
Voice (Australia)

Creative Director
Anthony De Leo

Designer
Anthony De Leo

BISS

Client
Bio Biss

Design Agency
StudioIN (Russia)

Design Director
Arthur Schreiber

Designer
Arthur Schreiber

Photography
Pavel Gubin

The graceful shape of the green glass bottle and the "tasty" label colors emphasize the naturalness of the beverages. The logo gives one an expectation of the pleasant hissing sound the lemonade makes when opened.

Our project included developing the name of the product, its positioning, the trademark, the shape of the bottle, the design of the label, and the variety of tastes. Our Art Director supervised the making of bottles and the printing of labels.

Epitome Late Harvest Riesling

The objective was to create a unique label for a sweet dessert wine. Target audience are females from the age of 25 and up.

The label takes its form from a doiley which is a reference to the types of food that are typically consumed with this variety of wine. The delicate pattern of the doiley creates a sense of elegance and quality that appeals to the target audience and visually sets the wine apart from the competition.

Client
Longview Vineyard

Design Agency
Voice (Australia)

Creative Director
Anthony De Leo, Scott Carslake, Shane Keane

Design Director
Anthony De Leo, Shane Keane

Designer
Shane Keane

Cognac FLEUR DE LIS

Femininity is characterized as the name and the elegance of lines and shapes. Using uncharacteristic of cognac capping is not only convenient but also practical. The design clearly distinguishes the product from the total mass of similar competitors.

Design Agency
StudioIN (Russia)

Design Director
Arthur Schreiber

Designer
Arthur Schreiber

Photography
Pavel Gubin

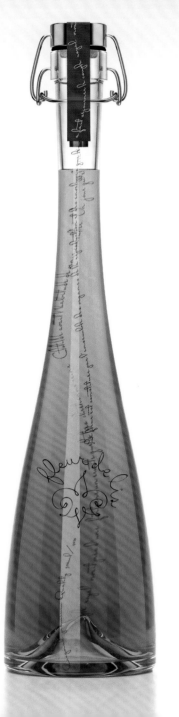

Åbro Ekologisk öl / Abro Organic Lager

Product design for Abro Organic Lager.

Client
AB Åbro Bryggeri / AB Åbro Breweries Sweden

Design Agency
Morris Pinewood Stockholm (Sweden)

Creative Director
Mattias Frodlund

Designer
Mattias Frodlund

1901 Red Ale

Bold City Brewery strives to link their brand of bold brew tightly to their beloved hometown of Jacksonville, Florida – the 1901 Red Ale is no exception. The complex and roasty flavored red ale is not only a favorite among patrons, but also commemorates Jacksonville's perseverance through the Great Fire of 1901 as one of the bold cities of the south.

Client
Bold City Brewery

Design Agency
Shepherd (USA)

Creative Director
Jacquie Wojcik

Design Director
Kendrick Kidd

Good Wine Company

Client
Good Wine Company

Design Agency
ilovedust (UK)

Brabante Beer

Brabante is a beer developed by a Spanish team of brewing masters. We provided new brand development, strategic positioning and design, and created corporate identity and package design for five different varieties.

Client
Brabante Cervezas S.L.

Design Agency
Tritone (Spain)

Creative Director
Dimas Gorostarzu

Designer
Dimas Gorostarzu

Photography
Antonio Mulero

S:t Eriks Pilsner

The story of S:t Erik's beer dates back to 1850, when S:t Erik's Brewery was established in Kungsholmstorg in central Stockholm, Sweden. The German-trained and highly skilled brew master Carl Magnus Peterson, who brewed beer between 1873 and 1905, made S:t Erik's beer quite popular in Stockholm quickly. In 2010 the classic St. Erik's Pilsner is re-launched with a new design and taste.

S:t Eriks Sommarlager

Client
Galatea Spirits

Design Agency
Entire Branding & Design (Sweden)

Creative Director
Mattias Brodén

Designer
Mattias Brodén

Photography
Pelle Lundberg

Ilustrator
Mattias Brodén & Giorgio Cantu

Black Drop

This is a packaging project about a premium dark beer, and is a job for the university. The concept of this design is the drop of water. The black drop represents the strong flavor of this beer. A simple yet powerful design.

Design Agency
Whoistoni (Spain)

Creative Director
Toni Garcia

Designer/Photography
Toni Garcia

Vodka VINTAGE

The design represents accurately the essence of the product. The vodka has an ingredient of grape, hence the shape of the bottle resembles the shape of grape.

Design Agency
StudioIN (Russia)

Design Director
Arthur Schreiber

Designer
Arthur Schreiber

Photography
Pavel Gubin

Samurai

The solution is to represent the word Samurai, thus there is a strict sword cut on the packing.

Design Agency
StudioIN (Russia)

Design Director
Arthur Schreiber

Designer
Arthur Schreiber

Photography
Pavel Gubin

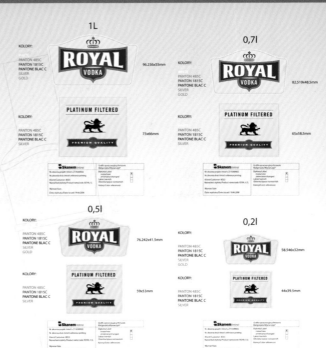

Royal Vodka

Bols Hungary is a dominant spirit company in the Hungarian market. The design of its vodka brand, Royal, needed to be refined – both the logo and the bottle itself. Still the designer had to keep the basic features of the original logo, e.g.: the red colour, the crown and lions.

The aim of this new design was to change the position of the brand to reach a new, younger target group. Keeping these conditions in mind the art director created a youthful, fresh bottle form and designed a new label featuring the traditional elements of the former one.

Client
Bols Hungary Ltd. - Bols Hungary Kft.

Design Agency
Café Design (Hungary)

Creative Director
Mr. Attila Simon

Design Director
Mr. Zsolt Kathi

40 islands

The design of the 40 Islands vodka is based on the idea of the lucidity and purity of the product and is presented by an illustration, the simplicity of the logo and the minimal area of the label.

The illustration gives one a feeling of a misty morning on a lake. The simplicity and austerity of the shape gives the product a higher status.

Client
PVVVK

Design Agency
StudioIN (Russia)

Design Director
Arthur Schreiber

Designer
Arthur Schreiber

Photography
Pavel Gubin

Vodka "Kacheli"

The company positions itself in the subprime segment, targeting the ad-sensitive consumers, for which the presence of an emotional component of a brand is highly important.

During the process of working on the project, several options were suggested to create a positive brand conception and strong brand identity. The solution suggested pulling the attention towards the natural products that were used in the making. This idea is supported by an eco-friendly, refined design, assuming the presence of a bright visual brand identity.

Client
"Raimbek-group" company

Design Agency
KIAN brand agency (Russia)

Creative Director
Kirill Konstantinov

Design Director
Andrey Kalashnikov

Designer
Igor Krivonogov

Index

Ah&Oh Studio
www.ahioh.com
P 043, 196-197

Anagrama
www.anagrama.com
P 068-071

Anti
www.anti-ink.com
P 026

Artilerya
www.artilerya.com
P 102

Baita Design Studio
www.baitadesign.com
P 018, 120

Base
www.basedesign.com
P 162-163, 186, 237

Bendita Gloria
www.benditagloria.com
P 240, 246-247

BIG FISH ®
www.bigfish.co.uk
P 078, 105, 115, 242

Bond. And SelectNY in New York
www.kaedingnyc.com
P 188, 202-203

Brandever
www.brandever.com
P 236

Bricault Design
www.bricault.ca
P 058-061

Bulldog
www.virtualbulldog.com
P 178

Café Design
www.cafedesign.hu
P 119, 171, 218-221, 235, 264-265

Camila Drozd, Freelance Studio
www.camiladrozd.com
P 044-045

CAMPBELL HAY
www.campbellhay.com
P 049, 210-211, 216

Chris Trivizas Design
www.christrivizas.eu
P 238-239

Christian Hanson
www.christianhanson.ca
P 048, 079, 104, 185

Comunicación Visual
www.calcco.com
P 224-225, 231

Cory Ingwersen
ingwersencory@gmail.com
P 166

Commando Group
www.commandogroup.no
P 172-173

CULDESAC
www.culdesac.es
P 134-135

Designkontoret Silver KB
www.silver.se
P 075, 114 152-153, 190-191, 206, 249

dfact
www.dfact.de
P 062-063

Doni & Associati
www.donieassociati.it
P 087, 092-095, 124-125, 136

Dow Design
eleanor@dowdeisgn.co.nz
P 022-023, 028

Emmaolivèstudio
www.emmaolivestudio.com
P 032

Entire Branding & Design
www.entire.se
P 027, 260

Estudio Clara Ezcurra
www.estudioclaraezcurra.com.ar
P 121, 126-127

Eulie Lee Design
www.eilue.com
P 194, 226

Face to face design
www.facetofacedesign.com
P 144

Ferroconcrete
www.ferro-concrete.com
P 029, 033

FFunction
www.ffctn.com
P 009

Flavia Oliveira
www.flaviaoliveiradesign.com
P 157

Fontos Graphic Design Studio
www.fgs.hu
P 006-007, 207

Hattomonkey
www.hattomonkey.ru
P 010-011, 025

Hoffman Creative
www.hoffman-creative.com
P 038-039, 042, 212-214

ilovedust
www.ilovedust.com
P 015, 122-123, 150, 160, 234, 258

Jennifer Cole Phillips
www.imptwitch.com
P 155

JJAAKK
www.jjaakk.com
P 064-065

KANELLA
www.kanella.com
P 074

Katie Hatz
www.katiehatz.com
P 054, 142, 182

KIAN brand agency
www.kian.ru
P 012-013, 035, 057, 101, 267

kissmiklos
www.kissmiklos.com
P 040

Kollor Design Agency
www.kollor.com
P 106

kopfloch.at
www.kopfloch.at
P 072, 145

L-ENFANT
www.l-enfant.com
P 174-177

Lavernia + Cienfuegos
www.lavernia.com
P 170, 193, 199

Liedgren Design
www.liedgrendesign.se
P 019

merkwuerdig.com
www.merkwuerdig.com
P 167

Milk
www.milk.se
P 083, 138-139

Mind Design
www.minddesign.co.uk
P 110-111, 158-159, 208-209

Morris Pinewood Stockholm
www.morrispinewood.se
P 222, 256

moruba
www.moruba.es
P 228, 243

MOUSE GRAPHICS
www.mousegraphics.gr
P 081, 107, 137, 169, 179

Moxie Sozo
www.moxiesozo.com
P 076-077, 128-129

Nadie Parshina studio
www.nadieparshina.ru
P 140, 241, 248, 250

NALINDESIGN
www.nalindesign.com
P 086

Neue Design Studio
www.neue.no
P 052-053

Neumeister Strategic Design
www.neumeister.se
P 021, 024, 156, 227, 233

NICO189
www.nico189.com
P 164-165

OCH
www.och-studio.se
P 215

ONLY Creatives
www.onlycreatives.com
P 189

Paprika
www.paprika.com
P 066-067

Pearlfisher
www. Pearlfisher.com
P 097, 118, 198

Peter Gregson Studio
www.petergregson.com
P 020, 041, 098, 103, 132

Pikartzo
www.pikartzo.com
P 192, 223

Policarpo Design
www.policarpodesign.com
P 084-085, 251

R Design
www.r-design.co.uk
P 187, 200

Reggs
www.reggs.com
P 055, 096, 201

Scott Lambert
www.design-positive.com
P 112-113

Shepherd
www.shepherd-agency.com
P 257

Sruli Recht
www.srulirecht.com
P 146-149

Steph Baxter
contactstephbaxter@gmail.com
P 034, 230

Stranger and Stranger Ltd
www.strangerandstranger.com
P 244-245

Strømme Throndsen Design
www.stdesign.no
P 030-031, 080, 091, 100, 116-117, 130-132

Studio 360
www.studio360.si
P 099

StudioIN
www.studioin.ru
P 014, 253, 255, 262-263, 266

Subconscious Co.,Ltd.
www.ssubconscious.blogspot.com
P 016-017

Taxi Studio
www.taxistudio.co.uk
P 082

The Creative Method
www.thecreativemethod.com
P 051, 184, 229

TIBOR+
www.tiborplus.de
P 008, 056

TomTor Studio
www.tomtor.com
P 036-037

Toykyo
www.toykyo.be
P 161, 183

Tritone
www.tritone.es
P 259

Turnstyle
www.turnstylestudio.com
P 108-109

Two Dot Two
www.twodotwo.com
P 180-181

Vanguard Works
www.vanguardworks.com
P 088-090

Vladimir Shmoylov
www.shmoylov.ru
P 154, 168

Voice
www.voicedesign.net
P 073, 232, 252, 254

We Are Plus
www.weareplus.com
P 204-205

WESEMUA
www.wesemua.com
P 133, 151

Whoistoni
www.whoistoni.com
P 261

Zeus Jones
www.designisfine.com
P 143

Zoo Studio
www.zoo.ad
P 046-047, 050, 195

DATE DUE	RETURNED
FEB 0 1 2012	JAN 3 1 2012
NOV 1 5 201	NOV 1 4 2012
NOV 0 7 2014	NOV 0 4 2014
OCT 0 3 2015	NOV 1 1 2015 SA